Food Authentication: Techniques, Trends and Emerging Approaches (Second Issue)

Food Authentication: Techniques, Trends and Emerging Approaches (Second Issue)

Editor

Raúl González-Domínguez

Basel • Beijing • Wuhan • Barcelona • Belgrade • Novi Sad • Cluj • Manchester

Editor
Raúl González-Domínguez
University of Huelva
Huelva, Spain

Editorial Office
MDPI
St. Alban-Anlage 66
4052 Basel, Switzerland

This is a reprint of articles from the Special Issue published online in the open access journal *Foods* (ISSN 2304-8158) (available at: https://www.mdpi.com/journal/foods/special_issues/Food_Authentication_Techniques_Trends_Emerging_Approaches_Second_Issue).

For citation purposes, cite each article independently as indicated on the article page online and as indicated below:

Lastname, A.A.; Lastname, B.B. Article Title. *Journal Name* **Year**, *Volume Number*, Page Range.

ISBN 978-3-0365-8920-6 (Hbk)
ISBN 978-3-0365-8921-3 (PDF)
doi.org/10.3390/books978-3-0365-8921-3

© 2023 by the authors. Articles in this book are Open Access and distributed under the Creative Commons Attribution (CC BY) license. The book as a whole is distributed by MDPI under the terms and conditions of the Creative Commons Attribution-NonCommercial-NoDerivs (CC BY-NC-ND) license.

Contents

About the Editor vii

Preface ix

Raúl González-Domínguez
Food Authentication: Techniques, Trends and Emerging Approaches (Second Issue)
Reprinted from: *Foods* **2022**, *11*, 1926, doi:10.3390/foods11131926 1

Valentina Fanelli, Isabella Mascio, Monica Marilena Miazzi, Michele Antonio Savoia, Claudio De Giovanni and Cinzia Montemurro
Molecular Approaches to Agri-Food Traceability and Authentication: An Updated Review
Reprinted from: *Foods* **2021**, *10*, 1644, doi:10.3390/foods10071644 5

Dirk W. Lachenmeier and Steffen Schwarz
Digested Civet Coffee Beans (Kopi Luwak)—An Unfortunate Trend in Specialty Coffee Caused by Mislabeling of *Coffea liberica*?
Reprinted from: *Foods* **2021**, *10*, 1329, doi:10.3390/foods10061329 25

Raúl González-Domínguez, Ana Sayago and Ángeles Fernández-Recamales
Potential of Ultraviolet-Visible Spectroscopy for the Differentiation of Spanish Vinegars According to the Geographical Origin and the Prediction of Their Functional Properties
Reprinted from: *Foods* **2021**, *10*, 1830, doi:10.3390/foods10081830 29

José Luis P. Calle, Marta Ferreiro-González, Ana Ruiz-Rodríguez, Gerardo F. Barbero, José Á. Álvarez, Miguel Palma and Jesús Ayuso
A Methodology Based on FT-IR Data Combined with Random Forest Model to Generate *Spectralprints* for the Characterization of High-Quality Vinegars
Reprinted from: *Foods* **2021**, *10*, 1411, doi:10.3390/foods10061411 41

Miso Kim, Junyoung Hong, Dongwon Lee, Sohyun Kim, Hyang Sook Chun, Yoon-Ho Cho, et al.
Discriminant Analysis of the Geographical Origin of Asian Red Pepper Powders Using Second-Derivative FT-IR Spectroscopy
Reprinted from: *Foods* **2021**, *10*, 1034, doi:10.3390/foods10051034 53

Siyu Yao, Didem Peren Aykas and Luis Rodriguez-Saona
Rapid Authentication of Potato Chip Oil by Vibrational Spectroscopy Combined with Pattern Recognition Analysis
Reprinted from: *Foods* **2021**, *10*, 42, doi:10.3390/foods10010042 67

Manuel León-Camacho and María del Carmen Pérez-Camino
SLE Single-Step Purification and HPLC Isolation Method for Sterols and Triterpenic Dialcohols Analysis from Olive Oil
Reprinted from: *Foods* **2021**, *10*, 2019, doi:10.3390/foods10092019 83

Nerea Núñez, Javier Saurina and Oscar Núñez
Authenticity Assessment and Fraud Quantitation of Coffee Adulterated with Chicory, Barley, and Flours by Untargeted HPLC-UV-FLD Fingerprinting and Chemometrics
Reprinted from: *Foods* **2021**, *10*, 840, doi:10.3390/foods10040840 97

Massimo Todaro, Vittorio Lo Presti, Alessandro Macaluso, Maria Alleri, Giuseppe Licitra and Vincenzo Chiofalo
Alkaline Phosphatase Survey in Pecorino Siciliano PDO Cheese
Reprinted from: *Foods* **2021**, *10*, 1648, doi:10.3390/foods10071648 111

Meei Chien Quek, Nyuk Ling Chin and Sheau Wei Tan
Optimum DNA Extraction Methods for Edible Bird's Nest Identification Using Simple Additive Weighting Technique
Reprinted from: *Foods* **2021**, *10*, 1086, doi:10.3390/foods10051086 . . . . **121**

About the Editor

Raúl González-Domínguez

Dr. Raúl González-Domínguez received his Ph.D. in Chemistry in 2015 (University of Huelva, Spain) and then moved to the University of Barcelona as a postdoctoral researcher. Currently, he holds a "Miguel Servet" excellence grant, funded by Instituto de Salud Carlos III, at "Instituto de Investigación e Innovación Biomédica de Cádiz" (INiBICA). His research interests mainly focus on the development of mass spectrometry-based metabolomics tools and chromatographic approaches for metabolite profiling, and their application in biomedical (e.g., age-related diseases and metabolic disorders), nutrition (e.g., the discovery of food intake biomarkers and the impact of diet on health), and food research (e.g., food authentication and traceability). He is the author of more than 95 research and review articles in peer-reviewed international journals, 13 book chapters, and 1 patent. He has participated in 19 international research projects and 25 Spanish national projects funded through competitive calls, as well as having held 7 R+D contracts with public and private entities. Dr. González-Domínguez has also been involved in the direction and supervision of more than 45 research projects for under- and post-graduate students.

Preface

The authentication of foods and beverages is a very current topic of great interest for all the actors involved in the food chain, including the food industry, consumers, and food science researchers. Food authenticity covers many different aspects related to mislabeling, adulteration, and misleading claims about origin, production methods, or processing technologies. As many factors may affect the chemical composition of foods (e.g., geographical origin, variety or breed, conditions of cultivation, and breeding and/or feeding), the implementation of accurate, robust, and high-throughput analytical methods is needed to assess their authenticity and traceability and, consequently, guarantee their safety and quality in terms of organoleptic, nutritional, and bioactive characteristics. For these purposes, multiple analytical tools can be employed in combination with advanced chemometrics, such as spectroscopic and chromatographic techniques, DNA-based methods, and state-of-the-art omics approaches. In this context, in 2020, the journal *Foods* launched the Special Issue "Food Authentication: Techniques, Trends and Emerging Approaches" to gather research papers and review articles dealing with the development and application of analytical techniques and emerging approaches in food authentication. Considering the success and popularity of this earlier Special Issue, we will now release a second Special Issue comprising ten valuable scientific contributions, including one review article, one commentary article, and eight original research articles.

Raúl González-Domínguez
Editor

Editorial

Food Authentication: Techniques, Trends and Emerging Approaches (Second Issue)

Raúl González-Domínguez [1,2]

[1] Agrifood Laboratory, Faculty of Experimental Sciences, University of Huelva, 21007 Huelva, Spain; raul.gonzalez@dqcm.uhu.es; Tel.: +34-959-219-975
[2] International Campus of Excellence CeiA3, University of Huelva, 21007 Huelva, Spain

The authentication of foods and beverages is a very current topic of great interest for all the actors involved in the food chain, including the food industry, consumers, and food science researchers. Food authenticity covers many different aspects related to mislabeling, adulteration, and misleading claims about origin, production method, or processing technologies. As many factors may affect the chemical composition of foods (e.g., geographical origin, variety or breed, conditions of cultivation, breeding and/or feeding), the implementation of accurate, robust, and high-throughput analytical methods is needed to assess their authenticity and traceability and, consequently, to guarantee their safety and quality in terms of organoleptic, nutritional, and bioactive characteristics. For these purposes, multiple analytical tools can be employed in combination with advanced chemometrics, such as spectroscopic and chromatographic techniques, DNA-based methods, and state-of-the-art omics approaches. In this context, the journal *Foods* launched the Special Issue "Food Authentication: Techniques, Trends and Emerging Approaches" in 2020 to gather research papers and review articles dealing with the development and application of analytical techniques and emerging approaches in food authentication [1]. Considering the success and popularity of this Special Issue, we now release a Second Issue comprising 10 valuable scientific contributions, including 1 review article, 1 commentary article, and 8 original research articles.

Fanelli et al. reviewed the most widely used DNA-based molecular techniques for authenticating and tracing fresh and processed agri-food products, from traditional molecular marker-based methods (e.g., single nucleotide polymorphisms) to more recent single region approaches (e.g., DNA barcoding, isothermal amplification-based methods) and next-generation-sequencing-based methods (e.g., DNA metabarcoding) [2]. Herein, an overview of recent advances and applications and an exhaustive comparison of the main advantages and limitations of each molecular method are provided. The importance of properly controlling the mislabeling and adulteration of digested coffees is reported in another commentary article [3]. The authors state that a great part of the coffee labelled as "Kopi Luwak" that can be found in the market is frequently adulterated with undigested coffee beans. Furthermore, they propose that the chemical and organoleptic characteristics of this specialty coffee could be majorly allocated to the diet of the civet cats (i.e., *Coffea* species, ripeness) rather than to changes caused by digestion.

Many of the original research articles published in this Special Issue revolve around the implementation of low-cost, ecofriendly, and non-destructive spectroscopic methods as a reliable alternative to traditional chemical-based analytical approaches for simple and rapid food authentication. In this respect, González-Domínguez et al. described the potential of ultraviolet-visible spectroscopy in combination with multivariate statistical tools to discriminate Spanish wine vinegars produced under three Protected Designations of Origin (PDO), namely, "Jerez", "Condado de Huelva", and "Montilla-Moriles" [4]. Additionally, regression analysis demonstrated that spectral data could accurately predict the physicochemical and functional properties of vinegars, particularly their total phenolic content

Citation: González-Domínguez, R. Food Authentication: Techniques, Trends and Emerging Approaches (Second Issue). *Foods* **2022**, *11*, 1926. https://doi.org/10.3390/foods11131926

Received: 23 June 2022
Accepted: 25 June 2022
Published: 28 June 2022

Publisher's Note: MDPI stays neutral with regard to jurisdictional claims in published maps and institutional affiliations.

Copyright: © 2022 by the author. Licensee MDPI, Basel, Switzerland. This article is an open access article distributed under the terms and conditions of the Creative Commons Attribution (CC BY) license (https://creativecommons.org/licenses/by/4.0/).

and antioxidant activity. Similarly, Fourier-transform infrared (FT-IR) spectroscopy was also found to be a practical methodology for the classification of Sherry vinegars according to their origin [5]. Statistical modelling was applied to develop a characteristic spectral fingerprint ("spectralprint") by selecting the most important variables, which enabled the rapid, reliable, and uncomplicated differentiation of vinegar samples depending on the starting wine. The same spectroscopic approach was employed in another study to discriminate Asian red pepper samples based on their geographical origin [6]. The four most significant peak variables from second-derivative FT-IR spectral data were selected as discriminant indicator variables, and their origin-specific ranges were set. These indicator ranges were able of successfully classifying all the samples under investigation. The last paper published in this Special Issue focused on the application of spectroscopic methods describes the utility of vibrational spectroscopy to predict the fatty acid profile of potato chips with the aim of authenticating the type of oil used in manufacturing [7]. Fatty acids were analyzed by gas chromatography with flame ionization detection (GC-FID), and spectral data were collected using Raman and near-infrared (NIR) sensors. Interestingly, pattern recognition analysis enabled the prediction of the major fatty acid composition and the detection of mislabeling issues.

As an alternative approach, other authors reported the use of chromatography-based techniques for authenticity and traceability purposes. León-Camacho and Pérez-Camino developed a new supported liquid extraction (SLE) method that, in combination with high-performance liquid chromatography (HPLC) and GC-FID, simplifies the isolation and quantification of the unsaponifiable fraction from fats and oils [8]. This procedure is easier, less time-consuming, and reduces the volume of solvents and reagents compared to traditional liquid–liquid extraction. Furthermore, this method ensured the efficient removal of fatty acids, thereby avoiding possible interferences during GC quantification and facilitating the determination of sterols and triterpenic dialcohols. In another study, untargeted fingerprinting analysis based on high-performance liquid chromatography with ultraviolet and fluorescence detection (HPLC-UV-FLD) was employed to detect common adulterants in coffee, namely, chicory, barley, and flours [9]. In combination with advanced chemometric tools, the methodology provided appropriate performance to detect and quantify adulterant levels down to 15% with good calibration and prediction errors.

The determination of alkaline phosphatase was also proposed as a potential marker for controlling cheeses produced under PDOs [10]. Alkaline phosphatase values in Pecorino Siciliano PDO samples were found to be strongly affected by the type of milk used during cheese production (i.e., raw milk vs. pasteurized milk) and by the temperature during cooking. This variability, probably because of the high craftsmanship, did not permit the researchers to establish clear ranges for discriminating cheeses depending on the production process. Alternatively, Quek et al. compared the overall performance of five DNA extraction procedures for identifying the origin of an edible bird's nest [11]. They concluded that a hybrid method, combining conventional SDS and a commercial kit (SDS/Qiagen), was the most suitable in terms of speed and cost-effectiveness.

In conclusion, the Special Issue "Food Authentication: Techniques, Trends and Emerging Approaches (Second Issue)" highlights the crucial importance of combining state-of-the-art analytical techniques with advanced statistical approaches with the aim of obtaining deeper insights into food composition and the discovery of novel authenticity indicators.

Funding: This research received no external funding.

Conflicts of Interest: The author declares no conflict of interest.

References

1. González-Domínguez, R. Food Authentication: Techniques, Trends and Emerging Approaches. *Foods* **2020**, *9*, 346. [CrossRef] [PubMed]
2. Fanelli, V.; Mascio, I.; Miazzi, M.M.; Savoia, M.A.; De Giovanni, C.; Montemurro, C. Molecular Approaches to Agri-Food Traceability and Authentication: An Updated Review. *Foods* **2021**, *10*, 1644. [CrossRef] [PubMed]

3. Lachenmeier, D.W.; Schwarz, S. Digested Civet Coffee Beans (Kopi Luwak)-An Unfortunate Trend in Specialty Coffee Caused by Mislabeling of Coffea liberica? *Foods* **2021**, *10*, 1329. [CrossRef] [PubMed]
4. González-Domínguez, R.; Sayago, A.; Fernández-Recamales, Á. Potential of Ultraviolet-Visible Spectroscopy for the Differentiation of Spanish Vinegars According to the Geographical Origin and the Prediction of Their Functional Properties. *Foods* **2021**, *10*, 1830. [CrossRef] [PubMed]
5. Calle, J.L.P.; Ferreiro-González, M.; Ruiz-Rodríguez, A.; Barbero, G.F.; Álvarez, J.Á.; Palma, M.; Ayuso, J. A Methodology Based on FT-IR Data Combined with Random Forest Model to Generate Spectralprints for the Characterization of High-Quality Vinegars. *Foods* **2021**, *10*, 1411. [CrossRef] [PubMed]
6. Kim, M.; Hong, J.; Lee, D.; Kim, S.; Chun, H.S.; Cho, Y.H.; Kim, B.H.; Ahn, S. Discriminant Analysis of the Geographical Origin of Asian Red Pepper Powders Using Second-Derivative FT-IR Spectroscopy. *Foods* **2021**, *10*, 1034. [CrossRef] [PubMed]
7. Yao, S.; Aykas, D.P.; Rodriguez-Saona, L. Rapid Authentication of Potato Chip Oil by Vibrational Spectroscopy Combined with Pattern Recognition Analysis. *Foods* **2020**, *10*, 42. [CrossRef] [PubMed]
8. León-Camacho, M.; Pérez-Camino, M.C. SLE Single-Step Purification and HPLC Isolation Method for Sterols and Triterpenic Dialcohols Analysis from Olive Oil. *Foods* **2021**, *10*, 2019. [CrossRef]
9. Núñez, N.; Saurina, J.; Núñez, O. Authenticity Assessment and Fraud Quantitation of Coffee Adulterated with Chicory, Barley, and Flours by Untargeted HPLC-UV-FLD Fingerprinting and Chemometrics. *Foods* **2021**, *10*, 840. [CrossRef]
10. Todaro, M.; Lo Presti, V.; Macaluso, A.; Alleri, M.; Licitra, G.; Chiofalo, V. Alkaline Phosphatase Survey in Pecorino Siciliano PDO Cheese. *Foods* **2021**, *10*, 1648. [CrossRef] [PubMed]
11. Quek, M.C.; Chin, N.L.; Tan, S.W. Optimum DNA Extraction Methods for Edible Bird's Nest Identification Using Simple Additive Weighting Technique. *Foods* **2021**, *10*, 1086. [CrossRef] [PubMed]

Review

Molecular Approaches to Agri-Food Traceability and Authentication: An Updated Review

Valentina Fanelli [1,*], Isabella Mascio [1], Monica Marilena Miazzi [1], Michele Antonio Savoia [1], Claudio De Giovanni [1] and Cinzia Montemurro [1,2,3]

1. Department of Soil, Plant and Food Sciences, University of Bari Aldo Moro, Via Amendola 165/A, 70126 Bari, Italy; mascioisa@gmail.com (I.M.); monicamarilena.miazzi@uniba.it (M.M.M.); michele.savoia@uniba.it (M.A.S.); claudio.degiovanni@uniba.it (C.D.G.); cinzia.montemurro@uniba.it (C.M.)
2. Spin off Sinagri s.r.l., University of Bari Aldo Moro, Via Amendola 165/A, 70126 Bari, Italy
3. Institute for Sustainable Plant Protection–Support Unit Bari, National Research Council of Italy (CNR), Via Amendola 122/D, 70126 Bari, Italy
* Correspondence: valentina.fanelli@uniba.it

Abstract: In the last decades, the demand for molecular tools for authenticating and tracing agri-food products has significantly increased. Food safety and quality have gained an increased interest for consumers, producers, and retailers, therefore, the availability of analytical methods for the determination of food authenticity and the detection of major adulterations takes on a fundamental role. Among the different molecular approaches, some techniques such as the molecular markers-based methods are well established, while some innovative approaches such as isothermal amplification-based methods and DNA metabarcoding have only recently found application in the agri-food sector. In this review, we provide an overview of the most widely used molecular techniques for fresh and processed agri-food authentication and traceability, showing their recent advances and applications and discussing their main advantages and limitations. The application of these techniques to agri-food traceability and authentication can contribute a great deal to the reassurance of consumers in terms of transparency and food safety and may allow producers and retailers to adequately promote their products.

Keywords: molecular traceability; authentication; agri-food; molecular markers; DNA barcoding; isothermal amplification; sequencing

1. Introduction

The major worries of consumers concern the origin and the safety of the food they buy. The increased awareness of the value of food quality induces the consumer to ask for transparency from food companies. At the same time, companies must be able to certify the content and origin of their products with the aim of protecting the consumer against fraud and adulterations. In this scenario, traceability and authentication are fundamental tools for reassuring consumers in terms of transparency and food safety and allowing producers to gain awareness of the value of their products. Traceability lets the tracking of the source of a food at any point in the production chain enabling the quality-control processes and cutting down the production of unsafe or poor-quality foods [1]. Food authentication is the process through which a food is tested to verify if it complies with the description contained in its label [2].

Traceability and authentication are integral components of the food safety and defense system and represent fundamental components of the food supply chain. A reliable authentication and traceability system can constitute an essential instrument for the protection of consumers, reducing the chance of people consuming adulterated or contaminated foods, and increasing supplier control and process safety. Consumers showed limited knowledge about the importance of authentication and traceability of food products [3,4], making essential the dissemination of the potential and reliability of tracing methods with the purpose to increase people's awareness of the role of food surveillance in health protection and the truthfulness of traceability information.

A wide variety of analytical methods for food traceability and authentication have been developed and tested [5]. Each method allows obtaining specific information on food composition and characteristics such as geographical origin, presence of adulterants, and species or varieties used in the production process. Among these analytical methods, the molecular approaches show some important advantages such as accuracy, sensitivity, and high reproducibility. Moreover, these methods are not affected by environmental changes, harvesting period, storage condition, and manufacturing process [6].

In the last decades, the demand for molecular tools for food authentication and traceability has significantly increased. This is mainly due to increasingly stringent legislation in the food sector and the market strategies aiming to assess a uniform and reliable control of the whole food chain from the field to the market and to ensure that consumer choices correspond to their expectations [7]. In this context, the European Union established two levels of recognition of food products: Protected Designation of Origin (PDO) and Protected Geographical Indication (PGI) with the purpose to protect the typical and local products and help consumers in choosing authentic food products and avoiding food frauds [8]. The DOP mark recognizes foods whose main characteristics depend on the territory of origin and the adherence to strict production rules. The IGP mark is attributed to a food that has a specific quality dependent on the specific geographical area of production. The availability of molecular analytical approaches is fundamental in the assessment of the conformity of PDO/PGI labels and the detection of not declared components.

Among the molecular analytical methods, some techniques such as the molecular markers-based approaches are well established, while some innovative approaches such as isothermal amplification-based methods and DNA metabarcoding have only recently found application in the surveillance of agri-products. Different authors have reviewed the most commonly used molecular methods for agri-food authentication [5,6,9–11], however, none of them have described the most recent and advanced techniques in detail and the potential of these methods in traceability and authentication processes.

In this review, an overview on the principal analytical methods for agri-food authentication and traceability was provided, focusing in particular on the molecular approaches. We describe some of the proven and widely tested molecular approaches such as molecular markers-based methods, showing their latest applications in agri-food surveillance. Moreover, we explore the most recent technologies describing their potential and prospects in food authentication and traceability. Finally, the advantages and limits of each approach are described and discussed.

2. Analytical Methods for the Traceability and Authentication of Food Deriving from Plant Species

In the last twenty years, an exponential growth of studies on methods for the traceability of animal- and plant-based food has been observed [12,13]. For animal-based food, the main frauds concern the substitution of an ingredient and the animal's geographical origin. In these cases, the analytical approaches are mainly based on vibrational spectroscopic techniques for the identification of the geographical origin and DNA typing of animal species [12]. For plant-based food, the fraudulent practices are highly disparate. The mismatch between product origin and geographical origin declared on the food label, the adulteration and contamination of product, the use of different species or different varieties compared with those declared on the label and the level of an additive higher than that permitted in a specific food are the most common frauds. Traceability approaches used for agri-foods are varied. Table 1 shows a list of the principal physico-chemical approaches used for plant-based food product traceability and authentication and the most recent reviews published for each method.

Table 1. Summary of the most recent reviews about the principal methods based on physico-chemical analysis for agri-food traceability and authentication and the food matrices on which they are commonly used.

Analytical Method	Food Products	References
Vibrational spectroscopic techniques	Different agri-food products	Lohumi et al. [14]
Mass spectrometry techniques	Different agri-food products	Castro-Puyana and Herrero [15]
Stable isotope analysis	Cereals, wine, and vegetable oils	Zhao et al. [16]
Gas chromatography coupled with mass spectrometry	Wine, hazelnuts, barley, terebinth, olive oil, coffee, vegetables, and fruits	Dymerski [17]
HPLC	Olive oil, coffee, tea, wine, juice, fruit, nuts	Esteki et al. [18]
Gas chromatography	Wine, chocolate, coffee, saffron, vegetable oil, fruit	Nolvachai et al. [19]
Spectroscopic and spectrometric techniques	Wine, vegetable oils, coffee, wheat, nuts, rice, vegetables and fruits	Medina et al. [20]
ELISA	Different agri-food products	Wu et al. [21]
Fluorescence spectroscopy	Vegetable oils, cereals, vegetables and fruits	Ahmad et al. [22]
Spectroscopic techniques	Vegetable oils, coffee, wine, fruit juice	Esteki et al. [23]
Raman spectroscopy	Olive oil, coffee, wine, rice	Xu et al. [24]
NMR	Balsamic vinegar, saffron, coffee, tomato	Consonni and Cagliani [25]
Mass spectrometry techniques	Wine, fruit juice, olive oil, beer, coffee	Rubert et al. [26]
Spectroscopic and spectrometric techniques	Different agri-food products	Wadood et al. [13]

Chromatography allows for the separation and quantification of macro- and micro-components in food products. The most widely used chromatographic techniques are high-performance liquid chromatography (HPLC) and gas chromatography. Both methods have been successfully used for the identification of the geographical origin of sweet cherry cultivars [27]. For agri-products, HPLC is an effective method to detect the presence of adulterants, quantify the level of additives, and identify the geographical origin of the product. The HPLC technique has been efficiently used for the authentication of extra virgin olive oil, the detection of adulteration in fruit juice, and the identification of the geographic origin of coffee, tea, and wine [18]. Gas chromatography is mostly applied in volatile substances analysis and detection of contaminants like pesticides. Gas chromatography analysis was performed to identify the geographical origin of different kinds of plant-based food products [13,19].

Immunoassays are analytical tools based on the use of antibodies or enzymes as recognition elements to detect the presence of specific antigens. Enzyme-linked immunosorbent assay (ELISA) is the most used immunological method for food traceability. This technique is mostly used for the detection of pesticide residues in food-borne matrices [28,29].

Spectroscopic techniques are fast and inexpensive methods based on the use of radiated energy to analyze the properties of a specific element. They have been widely used for different purposes including agri-food traceability. Fluorescence spectroscopy is a non-invasive and relatively inexpensive technique. However, it is less used compared to other spectroscopic methods due to its low detection limit. Despite this, fluorescence spectroscopy has been successfully used to detect adulteration in edible vegetable oils [30]. Vibrational spectroscopy is a widely used spectroscopic technique in the food sector. A wide array of vibrational spectroscopic methods including near-infrared (NIR), Fourier transform infrared (FTIR), and Raman spectroscopy have been used for the detection of adulteration and determining the authenticity of food products [14].

Nuclear magnetic resonance (NMR) allows for the identification of the composition of complex matrices of foodstuffs. The amount of any component in a mixture can be assessed with high precision. In the last years, NMR has been widely used for geographical traceability of agri-food products. This technique has been efficiently applied to the traceability of balsamic vinegar, saffron, coffee, and tomato [25], and recently to discriminate the origins of different species including rice, lentil, and citrus [31–33].

Among the most efficient methods for food authentication are the mass spectrometry (MS) techniques. A wide array of MS applications is available for food traceability and safety purposes such as the detection of contaminants, the composition, and the origin of a product [15]. Two MS techniques, isotope ratio mass spectrometry (IRMS), multi collector–inductively coupled plasma–mass spectrometry (MC-ICP-MS), are commonly used for the analysis of isotopic ratios in food matrices. The isotopic ratios are widely used in food authentication and traceability because they change with the area of origin of the product, climatic conditions, characteristics of soil, and agricultural practices. The most commonly used isotope ratios of elements for traceability of agri-products are $^{13}C/^{12}C$ and $^{15}N/^{14}N$, influenced by climate condition and agricultural practices; $^{2}H/^{1}H$ and $^{18}O/^{16}O$, affected by the area of origin; and $^{34}S/^{32}S$, influenced by geology [34]. Several studies have applied the analysis of isotopic ratios to identify the origin of agri-products [16].

Usually, the food traceability and authentication methods based on physico-chemical analysis are used in combination with each other in order to reach maximum sensitivity and reliability. The combined use of gas chromatography with mass spectrometry allows for accurate qualitative and quantitative analyses of complex mixtures providing noteworthy results in the surveillance of agri-products [17]. A recent study showed that the combined analysis of stable isotopes, elemental composition, and chemical markers was demonstrated to be highly effective in the determination of the geographical origin of a product [35].

Although over the years these analytical methods have been proven to be highly efficient and reliable in the identification of the geographical origin and potential adulterants fraudulently added to a product, they show remarkable limitations in the detection of contaminant species and in unmasking the use of varieties not declared in the product label. Additionally, physico-chemical approaches have been shown to be highly reliable with fresh products while they tend to lose effectiveness in the analysis of processed foods. These limitations are overcome by the use of molecular methods to food traceability.

3. Molecular Approaches to Agri-Food Analysis

DNA is a stable molecule present in all living organisms and each organism's DNA sequence is unique, enabling the distinguishing of the species and varieties used to produce a specific food. Moreover, DNA can also be recovered in enough quality and quantity in heavily processed food matrices. Thanks to the recent advancements in molecular biology and genetics, molecular approaches have become powerful and widely used methods for the authentication of agri-food products and for tracking the raw materials across the whole industry process. Along with the most widespread and experienced molecular marker-based methods, the more recent isothermal amplification-based methods, digital PCR techniques, and NGS-based approaches appear to be very promising in the traceability of a wide range of fresh and processed agri-foods. Table 2 shows a list of the most recent studies on agri-food authentication and traceability using DNA-based approaches.

Table 2. List of the most recent studies on DNA-based methods applied in the traceability and authentication of agri-foods.

Technique	Agri-Food Product	Detected Species	References
SSR/capillary electrophoresis	Grapes, must, and wine	Grapevine (*Vitis vinifera* L.)	[36]
SSR/HRM and SNP/HRM	Olive oil	Olive (*Olea europea* L.)	[37]
SSR/HRM	Olive oil	Olive (*Olea europea* L.)	[38]
SSR/capillary electrophoresis and SSR/HRM	Must and wine	Grapevine (*Vitis vinifera* L.)	[39]
SNP/PCR-RFLP	Olive oil	Olive (*Olea europea* L.)	[40]
SSR/capillary electrophoresis and SNP/Sanger sequencing	Extra virgin olive oil	Olive (*Olea europea* L.)	[41]
SNP/HRM	Must and wine	Grapevine (*Vitis vinifera* L.)	[42]
TaqMan SNP Genotyping Assay	Must and wine	Grapevine (*Vitis vinifera* L.)	[43]
SNP/biosensor	Must and wine	Grapevine (*Vitis vinifera* L.)	[44]
SNP/nanofluidic array	Coffee beans	Coffee (*Coffea arabica* L. and *Coffea canephora* Pierre ex. A. Froehner).	[45]
Species-specific primer PCR/sequencing	Flour, pasta, bread, and cookies	Common wheat (*Triticum aestivum* L.)	[46]
Species-specific primer /digital PCR	Flour and pasta	Common wheat (*Triticum aestivum* L.)	[47]

Table 2. Cont.

Technique	Agri-Food Product	Detected Species	References
Species-specific primer/digital PCR	Lotus seed paste	White kidney bean (*Phaseolus vulgaris* L.).	[48]
DNA barcoding/sequencing	Nutmeg mace	Nutmeg tree (*Myristica fragrans* Houtt)	[49]
DNA barcoding/capillary electrophoresis	Almond oil and almond kernels	Almond (*Prunus dulcis* Mill.)	[50]
Bar-HRM	Tea products	Tea (*Camellia sinensis* L.)	[51]
Bar-HRM	Nut species and walnut milk beverage	Walnut (*Juglans regia* L.), pecan (*Carya illinoensis* K. Koch), hickory (*Carya cathayensis* Sarg.), and peanut (*Arachis hypogaea* L.)	[52]
Bar-HRM	Raw seeds and ground coffee	Coffee (*Coffea arabica* L. and *Coffea canephora* Pierre ex. A. Froehner.)	[53]
DNA barcoding/NanoTracer strategy	Saffron powder	Saffron (*Crocus sativus* L.)	[54]
DNA barcoding/sequencing	Berry fruit and fruit juice	Different taxa	[55]
RPA-LFD	Saffron powder	Saffron (*Crocus sativus* L.)	[56]
DNA barcoding/LAMP	Saffron powder	Saffron (*Crocus sativus* L.)	[57]
LAMP	Durum wheat products	Durum wheat variety Aureo (*Triticum turgidum var. durum* L.)	[58]
DNA barcoding/LAMP	Fruit juice	Pomegranate (*Punica granatum* L.), Apple (*Malus domestica* (Suckow) Borkh.), and grape (*Vitis vinifera* L.)	[59]
Whole metagenome sequencing	Different herbal products	Different taxa	[60]
Whole metagenome sequencing	Lupin seed, flour, and cookies	Lupin (*Lupinus* spp.)	[61]
Whole chloroplast genome sequencing	Dried fruit	Different species of aromatic trees (*Zanthoxylum* spp.)	[62]
Whole chloroplast genome sequencing	Berry fruit	Different berry species (*Vaccinium* spp.)	[63]
DNA metabarcoding	Honey	Different taxa	[64]
DNA metabarcoding	Honey	Different taxa	[65]

3.1. Molecular Marker-Based Methods

Molecular marker-based methods are the most widely used techniques for food traceability. The main reasons are the reduced amount of template DNA required for marker detection, the chance to analyze simultaneously multiple target regions, and the possibility of obtaining both qualitative and quantitative information. In most cases, PCR-based methods are used to detect molecular marker variations [9]. PCR is diffusely employed in all molecular biology laboratories and does not require highly qualified personnel. Moreover, the low cost of the equipment and reagents makes PCR-based detection the easiest and most inexpensive method for molecular authentication and traceability of agri-products. The types of molecular markers most used for traceability purposes are microsatellite or Simple Sequence Repeat (SSR) and Single Nucleotide Polymorphism (SNP). They are highly informative due to their large number and even distribution throughout the genome and can highlight both inter and intra-species diversity [10].

3.1.1. Simple Sequence Repeats (SSR)

Over the last ten years, the number of works based on the use of SSR for agri-food traceability and authentication has progressively reduced, together with an increase in papers employing the more abundant and stable SNP markers (Figure 1), nevertheless, SSR remains the most widely used marker for molecular traceability. Simple sequence repeats are tandem repeated motifs of 2–6 bp flanked by highly conserved sequences. The polymorphism is due to the different number of repeats in the microsatellite region, and can be easily detected by PCR. Their high reproducibility and polymorphism degree make them a marker of choice for many applications including varietal identification and adulteration detection [66].

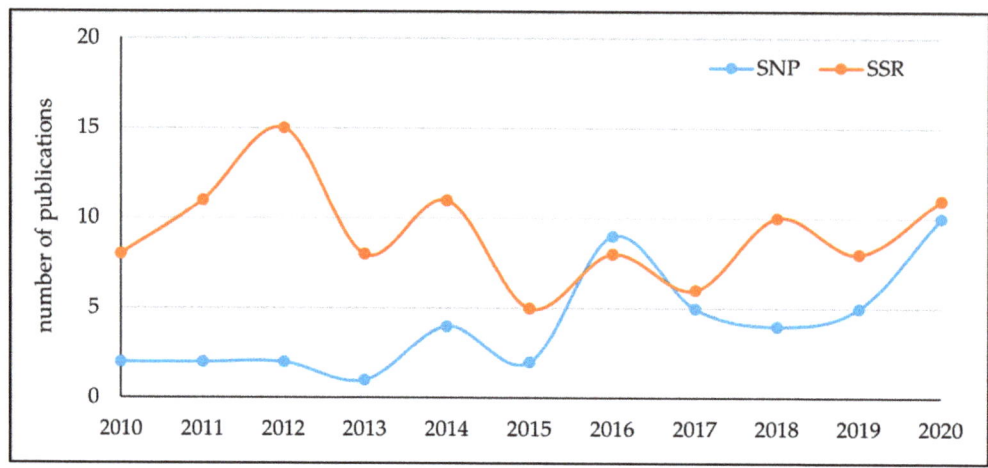

Figure 1. Number of publications per year in traceability and authentication of agri-foods through SSR and SNP markers. Data were obtained by searching the Scopus document archive (https://www.scopus.com; accessed on 8 June 2021) for English language articles for years between 2010 and 2020 using the following search terms: (SSR) AND (authentic*), (SSR) AND (traceability), (SNP) AND (authentic*) and (SNP) AND (traceability) and selecting only publications related to the agri-food sector and relevant to authentication and traceability processes.

Recently, SSRs have been efficiently used for the traceability of cocoa in beans and liquor [67], evaluations on trueness-to-type of raspberry [68] and olive [69] varieties, and to trace monovarietal and polyvarietal wines along the entire production chain [36]. Microsatellite markers have also been shown to be effective in tracing species characterized by a reduced diversity such as zucchini [70]. The most common approach involves the amplification of the regions of interest followed by fragment size evaluation through capillary electrophoresis. Nevertheless, the analysis of amplicons by the high resolution melting (HRM) assay was revealed to be highly effective in the authentication of PDO sweet cherry products [71] and the detection of adulteration in lentil [72]. Besides, the SSR-HRM technique allows for the authentication and traceability of processed food such as olive oil and wine. In particular, the combined use of SSR markers and HRM allows for distinguishing the varietal composition of olive oil and wine blends determining a limit of detection for adulteration included between 1% and 2.5% [37–39,73]. Moreover, microsatellite detection through real-time PCR enables the quantification of a specific contaminant. Pasqualone et al. [74] identified the common wheat contamination in durum wheat semolina and bread through the detection of genome D-specific SSR. The authors observed a detection limit of 3% and 5% for semolina and bread, respectively, by qualitative PCR lowered to 2.5% by real-time PCR.

3.1.2. Single Nucleotide Polymorphism (SNP)

Single nucleotide polymorphisms (SNPs) are variations in the DNA sequence involving a single base. They are the most abundant and ubiquitous markers in any living organism and their diallelic nature offers a lower error rate in allele calling compared with other molecular markers. Moreover, SNPs identification does not require DNA separation by size, and it is suitable for automation, making the analysis quick and reproducible.

SNPs are widely used in the traceability of animal-based foods, especially in the genetic authentication of meat [75], while only a few works are available in the agri-food sector, however, their use in this field has increased significantly in the last years and it is expected to keep growing in the future (Figure 1). The development of SNP-based approaches to agri-food traceability is encouraged by the increasing number of SNP catalogs mostly derived by GBS analysis. These panels are available in different species of agri-food interest such as grapevine, olive, pulses, cacao, and coffee [76–80]. Most of the works using a SNP-based traceability approach have focused on olive oil analysis [40,41], however, SNPs have been efficiently employed in differentiating Arabica and Robusta coffee varieties [81], in the authentication of Portuguese wine [42], and in the identification of Nebbiolo variety in musts and wines [43].

SNP identification is suitable for different detection methods such as single-base primer extension, cleaved amplified polymorphic sequences assays (CAPS), HRM, and sequencing techniques. Recently, an innovative system for wine authenticity based on the use of a biosensor as the system of SNP detection was developed by Barrias et al. [44]. DNA-based biosensors use DNA strands as probes for sensing DNA targets, distinguishing among samples differing for a single nucleotide in their sequence. The authors demonstrated the ability of the system to discriminate the varieties present in leaf, must, and wine samples, showing the promising application of this technique in SNP-based agri-food authenticity. Additionally noteworthy is the SNP genotyping system based on the use of a nanofluidic array. This system consists of the use of integrated fluidic circuits for high-throughput real-time PCR, allowing for the reliable analysis of multiple samples in a short time using small quantities of DNA. The nanofluidic SNP protocol has been successfully applied for cultivar authentication and identification of the adulterant varieties in cacao beans [82], discrimination of 40 tea varieties [83], and cultivar differentiation of coffee beans [45].

The rapid advances of next generation sequencing technologies have allowed for the automation of SNP detection, making the analysis based on this marker more rapid and reliable [84]. The employment of innovative sequencing approaches will allow the further spread of SNP-based approaches in the safeguarding of agri-food safety and quality.

3.2. Single Region Approaches

For some applications, the investigation focuses on a specific and well-known target DNA region. The analysis can be performed with the purpose to amplify a DNA sequence of a specific species or variety, taking advantage of peculiar differences in that region (e.g., indels). Conversely, PCR primers can be designed in a specific conserved region to amplify a sequence characterized by a certain polymorphism among species. This is the case of the DNA barcoding approach, representing an important tool for food traceability and authentication [85]. Isothermal amplification-based methods seem to be very promising and represent a novel group of nucleic acid amplification technologies that are simple and highly specific. Recently, these strategies have been successfully applied in the agri-food authentication sector.

3.2.1. Species-Specific Primer PCR

The presence of differences in nucleotide sequence or indels allows for the design of primers specific for a species or a variety. The detection of an amplification product makes possible the identification of adulterant species or variety in a particular food-borne sample. This approach has been widely used for the detection of common wheat in durum wheat-based products such as pasta or durum wheat bread. The identification of the presence of common wheat can be addressed by the detection of a sequence-specific of the D-genome, which is present in hexaploidy wheat but absent in durum wheat. Sonnante et al. [86] focused on the microsatellite region GDM111 to develop a quantitative method to detect the common wheat contamination in semolina, bread, and pasta products. The method was revealed to be effective up to a limit of 1% common wheat contamination. Matsuoka et al. [87] employed the *Starch Synthase II (SS II)* gene, coded on wheat A, B, and D genomes. The authors took advantage of some differences in the *SS II-D* gene to set up a quali-quantitative method for the detection of common wheat in blended flour. Silletti et al. [46] used a tubulin-based polymorphism to develop an assay specific for the detection of common wheat adulteration in pasta and flour. Through a DNA-based multiplex detection tool, Voorhuijzen et al. [88] were able to simultaneously test 15 different grain ingredients within one food with high accuracy.

In recent years, the development of techniques based on digital PCR (dPCR) has made the detection of a contaminant in food much faster and easier [89]. Digital polymerase chain reaction enables absolute quantification of a target nucleic acid in a sample even when the target is present at a very low number of copies. dPCR works by partitioning DNA fragments into thousands of independent droplets or chips, making it possible to directly count the number of target molecules through Poisson statistics [90]. dPCR has been widely used in the field of genetically modified organism (GMO) monitoring [91] and for pathogen diagnostics [92]. Moreover, this technique was also revealed to be very reliable and accurate in food safety and adulteration control. Pierboni et al. [93] efficiently applied droplet digital PCR to detect the presence of peanut and soybean allergens in mill and bakery products and demonstrated the usefulness of this technique for the food safety of allergic populations. More recently, Morcia et al. [47] developed a duplex chip digital PCR assay able to identify and quantify common wheat presence along the whole pasta production chain. The authors found that the limit of detection of the proposed method was 0.3% common wheat contamination, whereas the limit of quantification was found at the 1.5% level. Duplex droplet digital PCR and chip digital PCR were also revealed to be effective in the quantitative detection of kidney beans in lotus seed paste [48]. Generally, lotus seed paste is adulterated with cheaper ingredients such as common beans, making the detection method based on digital PCR extremely useful in revealing fraudulent substitutions or adventitious contaminations.

3.2.2. DNA Barcoding

DNA barcoding was developed by Hebert et al. [94] and is based on the analysis of variability within a specific genomic region called the "DNA barcode". This method represents an effective approach to food traceability and authenticity since it does not require extensive knowledge of the genome sequence of each organism and allows for the identification of more than one species at the same time. In animal-based food traceability, the barcoding is frequently based on the amplification of the *cytochrome oxidase* gene. In terrestrial plants, plastidial genes *rbcL* and *matK*, the trnH-psbA intergenic spacer and nuclear ITS2 sequence are mostly used as barcode regions [85]. DNA barcoding efficiency has been widely demonstrated in discriminating spices species such as nutmeg [49]. Recently, the analysis of trnH-psbA spacer and ITS2 sequence revealed them to be effective in the authentication of ginseng products [95] and the identification of adulterants in coffee and almond [50,96].

Frequently, DNA barcoding is employed coupled with high resolution melting (HRM) analysis (Bar-HRM). It consists in the amplification of a short DNA barcoding sequence and target region detection through HRM based on the distinctive melting behavior due to differences in DNA sequence. In the last years, the Bar-HRM strategy has found a large spread in agri-food surveillance. Bosmali et al. [97] set up a fast and cost-effective Bar-HRM method for PDO saffron authentication. The proposed approach was revealed to be highly effective in terms of specificity and sensitivity compared to other methods. A similar approach was used for the authentication of commercial sea buckthorn products [98]. More recently, Bar-HRM was employed for the authentication of several commercial tea products and detection of the presence of cashew DNA in the tea products [51], identification of common nut adulterants in walnut milk beverage [52], and the quantitative detection of Robusta traces in Arabica coffee products [53]. The great potential of the Bar-HRM technique has been widely demonstrated by Ballin et al. [99]. In this study, a DNA profiling platform for species authentication throughout the plant kingdom was developed through a multiplexed Bar-HRM approach. Distinct melting profiles were obtained for species originating from 29 different families spanning the angiosperms, gymnosperm, mosses, and liverwort, demonstrating the ability of the proposed approach in discriminating a large number of species without a priori knowledge of the species' DNA sequence.

DNA barcoding-based approaches in agri-food authentication and traceability are promising thanks to the great advances made in molecular biology techniques that allow us to combine the detection of a specific barcode sequence with modern technologies such as nanotechnologies. Based on this principle, Valentini et al. [54] developed an easy and inexpensive approach called "Nanotracer", which is able to detect the presence of a specific species-DNA in a food sample through a colorimetric response. The proposed approach is based on an asymmetric PCR amplification of a short barcode region, yielding a single-strand amplicon that is readily hybridizable to induce a color change due to the presence of DNA-functionalized gold nanoparticles. This method offers a rapid and naked-eye authentication test, and its implementation in the agri-food sector will provide an efficient system for food surveillance in the future.

The potential of the DNA barcoding strategy can be exploited through the sequencing of amplicons. The obtained sequence can be used to differentiate and univocally identify the species present in a food sample through a comparison with specific molecular databases. Recently, Sanger sequencing of specific DNA barcode regions was efficiently used for authentication of small berries in fruit products [55] and the construction of a DNA barcode library for the traceability of Chinese herbs [100]. However, the high costs and the limited number of samples that could be analyzed at the same time, along with the necessity of high-quality DNA, led Sanger sequencing to be supplanted by the next generation sequencing (NGS) technologies, which offer a much higher throughput through a less expensive and less time-consuming procedure.

The adoption of a universal barcode shows evident limits at the cultivar level, where genetic variability is limited. To overcome these limits, the ultra-barcoding methodology was proposed [101] to obtain a varietal identification. This strategy is based on the sequencing of the whole plastidial genome and a portion of the nuclear genome through NGS technologies. Ultra-barcoding has been shown to be a highly reliable strategy in cacao authentication [102].

The use of the DNA barcoding method in the agri-food sector is supported by the availability of the Barcode of Life Database (BOLD) coordinated by the International Barcode of Life Project [103]. This database contains a reference library for all living species, allowing the identification of more than 300,000 species on the base of the barcode sequence. Moreover, it includes a comprehensive registry of primers useful in the generation of barcode sequences. BOLD is a reliable resource for the exploitation of the potentiality of the DNA barcoding approach in food authenticity and safety.

3.2.3. Isothermal Amplification-Based Methods

Isothermal amplification-based techniques represent a promising alternative to classical PCR since they achieve rapid and efficient detection of a nucleic acid target without requiring the use of a thermocycler. These methods allow the amplification of a specific region in an exponential manner at a constant temperature. Over the last decade, various techniques based on isothermal amplification have been developed; although their features can vary among the different methods, they share some characteristics such as the use of a polymerase with strand-displacement activity. Some of the isothermal amplification techniques mostly used in agri-food surveillance are rolling circle amplification (RCA), multiple displacement amplification (MDA), recombinase polymerase amplification (RPA), and loop-mediated isothermal amplification (LAMP). These methods are mostly used in the detection of various micro-organisms, representing an important instrument to control food-borne diseases and safeguard food safety and quality [104]. Furthermore, they were also revealed to be highly sensitive and efficient in agri-food authentication and traceability. RPA in combination with ELISA has been shown to be highly effective in the detection of allergens such as hazelnut, peanut, and soybean as well as undeclared food ingredients [105]. Recently, Zhao et al. [56] proposed a novel analysis based on the combined use of RPA and lateral flow device (RPA-LFD) for saffron authentication. This rapid assay was revealed to be highly sensitive and specific, with no cross-reaction with common saffron adulterants.

Among the isothermal amplification-based methods, LAMP is the most widely used. This technique employs four to six different primers able to recognize six to eight different sequences of a target region, allowing the synthesis of large amounts of DNA in a short time. The amplification products are stem-loop DNAs with different inverted target repeats; these products can be detected with different methods including real-time assay and naked-eye detection through DNA-binding dyes or colorimetric indicators [106]. The high specificity, efficiency, and simplicity of the LAMP method has led to its application in the identification of different micro-organisms including food-related pathogens [107]. This approach is also suitable for the detection of GMOs through the employment of commonly used promoters or marker genes as LAMP targets [108]. Recently, LAMP has also assumed a relevant role in agri-food surveillance for the identification of specific species or even a variety in a specific food product. This approach has been used to authenticate saffron and discover its adulterants such as safflower and turmeric [57]. Cibecchini et al. [58] set up a portable colorimetric LAMP-based method to detect the presence of a specific wheat variety (Aureo) in grains and flours. Hu and Lu [59] developed a device for the specific detection of pomegranate, apple, and grape DNA present in fresh fruit juice. The authors combined DNA extraction and LAMP reaction in a hybrid paper/polymer-based lab-on-a-chip platform, allowing for the quick detection of a specific species in a juice sample through the use of a fluorescent dye. In the future, this method is expected to play an important role in the field of agri-food authentication and traceability.

3.3. Next Generation Sequencing-Based Methods

DNA sequencing represents the easiest way to detect multiple species and varieties present in a specific food-borne sample. Traditional Sanger sequencing allows for the detection of a specific DNA region at a time. Although cloning may improve resolution, it requires numerous steps and is very time-consuming. Moreover, Sanger sequencing is a relatively slow method, producing reads with a length not exceeding 900 bp [109]. Next generation sequencing (NGS) is a high throughput technique enabling the generation of different quantities and lengths of DNA sequencing. The different approaches are commonly grouped based on the length of reads produced during the sequencing. Therefore, we distinguished between short-read and long-read sequencing methods defined as second- and third-generation technologies, respectively.

The short-read sequencing approaches such as sequencing by synthesis and ion semiconductor sequencing were the first NGS techniques to be developed. Illumina is the current leader for the short-read sequencing approach. This technique is based on the peculiar bridge amplification method and the sequencing by synthesis strategy, which generates long-reads up to 300 bp [110]. Another popular short-read strategy is the ion semiconductor sequencer Ion Torrent based on the use of a dedicated sensor that acts as a highly sensitive pH meter, which detects the hydrogen ion release associated with nucleotide incorporation into the growing strand. For authentication of processed foods, the short-read-based sequencing strategies are preferable since DNA recovered from these matrices is usually highly degraded.

Third-generation strategies are quite recent techniques that enable overcoming many of the limitations of short-read sequencing through the sequencing of a single DNA/RNA molecule and generating reads with a length between 1 kb and 2 Mb [110]. The main long-read approaches are the single-molecule real-time sequencing (SMRT) and the nanopore sequencing. Despite the great potential of these techniques, their use is extremely limited in the food traceability sector.

Although the use of NGS technologies has spread in several diagnostics and research sectors, their use in the field of agri-food molecular traceability remains limited. A possible explanation is that NGS technologies present high costs and require extensive computational power. In addition, these strategies require high-quality DNA, which is not always possible to recover from highly processed foods. Nevertheless, a certain number of studies on agri-food traceability and authentication through NGS-based approaches have been published. There are basically two adopted strategies: whole metagenome sequencing and DNA metabarcoding.

3.3.1. Whole Metagenome Sequencing

Whole metagenome sequencing (WMS) allows scanning for several species simultaneously even when these are present in a small quantity in a food matrix [111]. This approach is widely used in the food security sector to identify and characterize complex microbial communities in food samples [112]. An important advantage of using WMS in food-borne hurtful microbial detection is the possibility of also detecting non-culturable pathogens; moreover, the production of draft genome sequences of the bacteria responsible for food-borne alerts is also possible, allowing for the identification of contamination sources [113]. Likewise, WMS can be employed to trace specific species and even varieties with very high sensitivity and specificity. The analysis of whole genomes allows for the authentication and detection of non-approved species. Complex food matrices can be analyzed, and the detected reads assigned to corresponding organisms by comparison with "ad hoc" databases.

A software pipeline, called AFS (All-Food-Seq), was developed to quantitatively measure the species composition in food-borne samples. This pipeline takes advantage of the deep sequencing of total DNA, allowing for the identification of species components through the mapping of reads to publicly available reference genome sequences and the quantification of species proportions based on a sequence read counting approach. This method has been successfully applied for the traceability and authentication of different animal- and plant-based foods [111].

More recently, Haiminen et al. [114] set up a bioinformatic pipeline, FASER (Food Authentication from SEquencing Reads), to resolve the relative composition of mixtures of eukaryotic species using RNA or DNA sequencing. Moreover, they developed a comprehensive database including more than 6000 plants and animals that may be present in food. FASER was revealed to be a highly sensitive and accurate method to detect fraudulent substitutions or contaminations in the most disparate food matrices.

Whole metagenome sequencing has been proved to be very effective in the identification and authentication of herbal products [60] and the detection of contaminants in food processed samples [61]. In the latter work, the authors combined metagenomic sequencing and an alignment-free k-mer based approach for the identification of plant DNA in processed samples. In particular, they demonstrated that lupin DNA can be individuated in controlled mixtures of sequences from the target and closely related non-target species, showing that lupin-specific components are detectable in baked cookies containing a minimum of 0.05% of lupin flour in wheat flour.

The whole chloroplast genome can be sequenced as an alternative to nuclear DNA for food authentication purposes. This is particularly useful in highly processed agri-foods since organellar DNA is present in high copy numbers compared to nuclear DNA, preventing degradation occurring during the production process. The sequencing of chloroplast genome produces reads that can be compared to specific databases containing complete chloroplast genome sequences such as the GenomeTrakrCP, which is publicly available at the National Center for Biotechnology Information (https://www.ncbi.nlm.nih.gov/bioproject/PRJNA325670/; accessed on 24 May 2021) [115]. This approach has been demonstrated to be highly effective by several authors [62,63].

3.3.2. DNA Metabarcoding

The DNA metabarcoding approach combines the high throughput sequencing strategies with DNA barcoding, allowing the analysis of multiple amplicons corresponding to different barcode regions by sequencing them in parallel. The general strategy is based on extracting the whole DNA from certain foods, amplifying a specific barcode region whose dimensions can vary from 120 up to 600 bp, sequencing the corresponding amplicon, and analyzing the sequence using specific pipelines. This strategy is particularly suitable for highly processed foods since the DNA extracted from these matrices is usually degraded, making possible only the amplification of short regions [7]. Moreover, the DNA metabarcoding approach has also been demonstrated to be useful for quantitative analysis. In fact, differences in sequence reads abundance between species can be used to infer the corresponding differences in species abundance in a food sample [116].

The most commonly used plant barcode regions for DNA metabarcoding analysis are the nuclear ITS regions or the plastidial rbcL and psbA-trnH. In particular, the ITS1 and ITS2 regions have been used to identify plant components in herbal teas through their sequencing through two different platforms, Illumina and Ion Torrent, showing that both sequencing strategies are effective in qualitative and quantitative detection of different species [117]. Frigerio et al. [118] analyzed the sequence variability at DNA barcoding psbA-trnH and ITS and minibarcoding rbcL 1-B regions to trace medicinal and aromatic plants. Recently, a comprehensive ITS reference dataset called PLANiTS including all the ITS sequences of the Viridiplantae clade was developed [119]. The PLANiTS dataset represents a reliable first step toward an accurate standardization of plant DNA metabarcoding studies.

The effectiveness of DNA metabarcoding in the agri-food authentication and traceability sector has been widely demonstrated in the authentication of polyfloral and monofloral honey [64,65,120]. In these cases, the metabarcoding approach allowed not only for the identification of the botanical composition of honey, but also to investigate its geographical origin based on the genetic characterization of pollen content.

Recently, Gostel et al. [121] developed microfluidic enrichment barcoding (MEBarcoding) for high-throughput plant barcoding, a cost-effective method based on the combined use of the Fluidigm Access Array and Illumina MiSeq. This study enabled them to build a highly comprehensive barcode database and demonstrated that the proposed approach is efficient in discriminating a very large number of species present in a food-borne matrix at the same time.

4. Advantages and Limits of Molecular Methods in Agri-Food Authentication and Traceability

A wide variety of analytical techniques for authentication and traceability of agri-food products have been developed and tested. For a long time, chemical and biochemical approaches have been used for the detection of specific components in foodstuffs; nevertheless, in the last few decades, molecular techniques have taken the upper hand in the food surveillance sector. DNA-based methods are mostly used for the identification and quantification of species and varieties composed of fresh or processed food. Indeed, DNA is present in nearly all the cells of a given organism and its sequence remains unchanged during all production phases. Instead, proteins and secondary metabolites may be influenced by growing conditions, harvesting period, and storage environment [6]. Moreover, DNA is a much more resistant molecule to industrial transformation compared to other biological components. On the other hand, physical fragmentation and chemical treatment can affect the yield, integrity, and quality of DNA [11]. For this reason, several protocols for DNA extraction from processed agri-food matrices were developed with the aim to recover a sufficient amount of good-quality DNA for subsequent analysis (Table 3). These protocols were optimized to extract DNA from a specific food-borne product with the purpose of maximizing the yield while minimizing the coextraction of enzymatic reaction inhibitors.

Table 3. List of the most recent protocols for DNA extraction from processed agri-foods and related references.

Agri-Food Matrices	Method	Reference
Must and wine	CTAB-based method/post-extraction purification	di Rienzo et al. [39]
Extra virgin olive oil	Hexane-based method	Piarulli et al. [41]
Nutmeg mace	SDS-based method	Swetha et al. [49]
Fruit juice	Filtration device	Hu and Lu [59]
Soybean oil	CTAB-based method	Xia et al. [122]
Wine	CTAB-based method	Pereira et al. [123]
Groundnut oil	DNA enrichment/CTAB-based method	Bojang et al. [124]
Honey	CTAB-based method	Soares et al. [125]
Sesame and flaxseed	SDS-based method/post-extraction purification	López-Calleja et al. [126]

A valid alternative to nuclear DNA-based analysis is the use of approaches involving the chloroplast genome, which is present in high copy numbers in vegetal cells. Indeed, heavily industrial treatments can severely affect nuclear DNA quality and quantity, while this occurs to a lesser extent with chloroplast DNA due to its abundance [62,63].

Despite the significant advances that have been made in molecular techniques, innovative approaches are only partially used in agri-food authentication, while traditional molecular marker-based methods, whose effectiveness have been amply demonstrated, remain the approaches of choice. Regarding molecular marker-based methods, SNPs and SSRs are largely used nowadays because of their standardized and straightforward detection systems. These approaches are used mainly in the identification of plant varieties aiming to prevent fraudulent commercial activities. SNP and SSR application for food traceability and authentication offer several advantages: they have a high level of polymorphism, high reproducibility, and can be detected on a very small portion of DNA, which in the case of fragmented DNA may constitute an important advantage [127]. Moreover, recent technical advances in SNP detection have made this marker an election tool in food traceability. Indeed, modern sequencing technologies allow millions of SNPs to be processed, simultaneously making possible the analysis of several samples in extremely short times [128]. Nevertheless, being highly species-specific, the molecular marker-based methods require the knowledge of plant species putatively present in a food and access to the correct DNA sequence of interest. Therefore, their application is often limited to a single species [129].

Frequently, a food can contain several vegetal species and the availability of an instrument able to detect all the species simultaneously becomes necessary for traceability and authentication purposes. Approaches based on DNA barcoding represent an effective alternative to DNA fingerprinting methods in plant identification since they do not require the knowledge of the whole genome of an organism, being based on the exploitation of one or few genomic regions [11]. DNA barcoding shows two important advantages: the requirement to amplify a very short DNA region (a few hundred base pairs) and the widespread use of plastidial genome, which is more preserved during industrial processing [85]. Moreover, the availability of several plant DNA barcoding databases considerably simplifies species detection and identification [103,119,121]. Nevertheless, DNA barcoding presents some important limitations. First, only the species for which a reference is available can be identified; therefore, database incompleteness greatly affects the reliability of analysis [109]. Another important limit of DNA barcoding is that it can only be applied to identify monophyletic species, since polyphyletic and paraphyletic species do not display a clear barcode gap (i.e., a gap between frequency distributions between intra- and interspecific distances). The absence of a barcoding gap makes the definition of a threshold value to identify species impossible, generating either false negatives (species missed) or false positives (false species) [130]. This consideration makes evident the limitations of adopting a barcode-based strategy for cultivar distinction. Therefore, in some cases, a combined approach of molecular markers and DNA barcode would be the best strategy for an accurate and exhaustive authentication analysis [72,97].

Whole metagenome sequencing is the best strategy for authenticity, since it allows for the detection of additives, poisonous plants, allergens, and any other kind of adulterants fraudulently or accidentally added to a food-borne product. The main limitation of NGS-based methods in agri-food authentication is the obtainment of sufficiently high-quality DNA. This step is crucial to ensure that all DNA sequences present in a food-borne sample are properly identified [113]. A large number of DNA extraction protocols are now available for different kinds of foods including highly processed products (Table 3). These protocols take into account the specific features of a product implementing a series of steps aiming at the collection of a minimum amount of sufficient quality DNA on one hand, and the removal of inhibitors on the other. In some cases, the tuned protocol resulted in being highly effective in isolating DNA suitable for high throughput approaches [65,120]. Despite the great potential, the current use of NGS within the agri-food authentication and traceability sector is limited compared to the more established techniques. In the near future, the technological advances of NGS techniques, along with a cost reduction and more user-friendly options for analysis, will make these approaches increasingly widespread in food authenticity.

5. Conclusions

Agri-food traceability and authentication require reliable and accurate methods for the identification of plant species and varieties in a wide collection of fresh and processed food, without ambiguity. The possibility of being aware of the composition of a food has assumed increasing importance among consumers, thanks to the action of mass communication concerning the relevance "of knowing what one is eating". Among the different traceability techniques, molecular approaches are gaining increasing interest due to their significant advantages compared to the physico-chemical approaches.

There are many various molecular methods suitable for agri-food surveillance. Some of them such as the molecular marker-based approaches have been extensively experienced and used in the agri-food sector; several authors have described their main applications in detail. Here, we presented the advances of these approaches and their most recent employment in agri-food traceability and authentication. Moreover, we provided an extensive description of the most innovative approaches such as isothermal amplification-based methods and DNA metabarcoding, which have only recently found application in agri-food surveillance. We highlighted their potential and prospects by showing the latest works on traceability and authentication based on the use of these methods. Finally, the description of the main advantages and limits of each molecular method will represent an effective prompt for anyone who wants to find the best method to authenticate or trace a specific agri-food.

The wide panel of molecular techniques to traceability and authentication in the agri-food sector constitutes a powerful tool to protect both producers and consumers, ensuring consumer freedom of choice and improving the transparency of food production systems, therefore allowing honest producers to adequately promote their food products.

Author Contributions: Conceptualization, V.F. and C.M.; Methodology, V.F., I.M., and M.M.M.; Data curation, V.F., M.A.S., and C.D.G.; Writing—original draft preparation, V.F.; Writing—review and editing, V.F., I.M., M.M.M., M.A.S., C.D.G., and C.M.; Supervision, C.M. All authors have read and agreed to the published version of the manuscript.

Funding: This research received no external funding. The APC was funded by MIUR-PON Ricerca e Innovazione 2014–2020 (project AIM1809249-attività 2, linea 1).

Conflicts of Interest: The authors declare no conflict of interest.

List of Abbreviations

HPLC	High-performance liquid chromatography
ELISA	Enzyme-linked immunosorbent assay
NMR	Nuclear magnetic resonance
SSR	Simple sequence repeat
SNP	Single nucleotide polymorphism
HRM	High resolution melting
RFLP	Restriction fragment length polymorphism
RPA	Recombinase polymerase amplification
LFD	Lateral flow device
LAMP	Loop-mediated isothermal amplification
GMO	Genetically modified organism
PDO	Protected designation of origin
PGI	Protected geographical indication
NGS	Next generation sequencing
WMS	Whole metagenome sequencing

References

1. Aung, M.M.; Chang, Y.S. Traceability in a food supply chain: Safety and quality perspectives. *Food Control* **2014**, *39*, 172–184. [CrossRef]
2. Ballin, N.Z.; Laursen, K.H. To target or not to target? Definitions and nomenclature for targeted versus non-targeted analytical food authentication. *Trends Food Sci. Technol.* **2019**, *86*, 537–543. [CrossRef]
3. Maitiniyazi, S.; Canavari, M. Exploring Chinese consumers' attitudes toward traceable dairy products: A focus group study. *J. Dairy Sci.* **2020**, *103*, 11257–11267. [CrossRef] [PubMed]
4. Pelegrino, B.O.; Silva, R.; Guimarães, J.T.; Coutinho, N.F.; Pimentel, T.C.; Castro, B.G.; Freitas, M.Q.; Esmerino, E.A.; Sant'Ana, A.S.; Silva, M.C.; et al. Traceability: Perception and attitudes of artisanal cheese producers in Brazil. *J. Dairy Sci.* **2020**, *103*, 4874–4879. [CrossRef]
5. Danezis, G.P.; Tsagkaris, A.S.; Camin, F.; Brusic, V.; Georgiou, C.A. Food authentication: Techniques, trends & emerging approaches. *Trends Analyt. Chem.* **2016**, *85*, 123–132.
6. Lo, Y.T.; Shaw, P.C. DNA-based techniques for authentication of processed food and food supplements. *Food Chem.* **2018**, *240*, 767–774. [CrossRef]
7. Galimberti, A.; Casiraghi, M.; Bruni, I.; Guzzetti, L.; Cortis, P.; Berterame, N.M.; Labra, M. From DNA barcoding to personalized nutrition: The evolution of food traceability. *Curr. Opin. Food Sci.* **2019**, *28*, 41–48. [CrossRef]
8. Regulation (EU) No 1151/2012 of the European Parliament and of the Council of 21 November 2012 on Quality Schemes for Agricultural Products and Foodstuffs. Available online: https://eur-lex.europa.eu (accessed on 9 July 2021).
9. Martins-Lopes, P.; Gomes, S.; Pereira, L.; Guedes-Pinto, H. Molecular markers for food traceability. *Food Technol. Biotechnol.* **2013**, *51*, 198–207.
10. Scarano, D.; Rao, R. DNA markers for food products authentication. *Diversity* **2014**, *6*, 579–596. [CrossRef]
11. Corrado, G. Advances in DNA typing in the agro-food supply chain. *Trends Food Sci. Technol.* **2016**, *52*, 80–89. [CrossRef]
12. Abbas, O.; Zadravec, M.; Baeten, V.; Mikuš, T.; Lešić, T.; Vulić, A.; Prpić, J.; Jemeršić, L.; Pleadin, J. Analytical methods used for the authentication of food of animal origin. *Food Chem.* **2018**, *246*, 6–17. [CrossRef] [PubMed]
13. Wadood, S.A.; Boli, G.; Xiaowen, Z.; Hussain, I.; Yimin, W. Recent development in the application of analytical techniques for the traceability and authenticity of food of plant origin. *Microchem. J.* **2020**, *152*, 104295. [CrossRef]
14. Lohumi, S.; Lee, S.; Lee, H.; Cho, B.K. A review of vibrational spectroscopic techniques for the detection of food authenticity and adulteration. *Trends Food Sci. Technol.* **2015**, *46*, 85–98. [CrossRef]
15. Castro-Puyana, M.; Herrero, M. Metabolomics approaches based on mass spectrometry for food safety, quality and traceability. *Trends Analyt. Chem.* **2013**, *52*, 74–87. [CrossRef]
16. Zhao, Y.; Zhang, B.; Chen, G.; Chen, A.; Yang, S.; Ye, Z. Recent developments in application of stable isotope analysis on agro-product authenticity and traceability. *Food Chem.* **2014**, *145*, 300–305. [CrossRef] [PubMed]
17. Dymerski, T. Two-dimensional gas chromatography coupled with mass spectrometry in food analysis. *Crit. Rev. Anal. Chem.* **2018**, *48*, 252–278. [CrossRef]
18. Esteki, M.; Shahsavari, Z.; Simal-Gandara, J. Food identification by high performance liquid chromatography fingerprinting and mathematical processing. *Food Res. Int.* **2019**, *122*, 303–317. [CrossRef]
19. Nolvachai, Y.; Kulsing, C.; Marriott, P.J. Multidimensional gas chromatography in food analysis. *Trends Analyt. Chem.* **2017**, *96*, 124–137. [CrossRef]
20. Medina, S.; Perestrelo, R.; Silva, P.; Pereira, J.A.M.; Câmara, J.S. Current trends and recent advances on food authenticity technologies and chemometric approaches. *Trends Food Sci. Technol.* **2019**, *85*, 163–176. [CrossRef]

21. Wu, L.; Li, G.; Xu, X.; Zhu, L.; Huang, R.; Chen, X. Application of nano-ELISA in food analysis: Recent advances and challenges. *Trends Analyt. Chem.* **2019**, *113*, 140–156. [CrossRef]
22. Ahmad, M.H.; Sahar, A.; Hitzmann, B. Fluorescence spectroscopy for the monitoring of food processes. In *Measurement, Modeling and Automation in Advanced Food Processing*; Hitzmann, B., Ed.; Springer: New York, NY, USA, 2017; Volume 161, pp. 121–151.
23. Esteki, M.; Shahsavari, Z.; Simal-Gandara, J. Use of spectroscopic methods in combination with linear discriminant analysis for authentication of food products. *Food Control* **2018**, *91*, 100–112. [CrossRef]
24. Xu, Y.; Zhong, P.; Jiang, A.; Shen, X.; Li, X.; Xu, Z.; Shen, Y.; Sun, Y.; Lei, H. Raman spectroscopy coupled with chemometrics for food authentication: A review. *Trends Analyt. Chem.* **2020**, *131*, 116017. [CrossRef]
25. Consonni, R.; Cagliani, L.R. The potentiality of NMR-based metabolomics in food science and food authentication assessment. *Magn Reson Chem.* **2019**, *57*, 558–578. [CrossRef]
26. Rubert, J.; Zachariasova, M.; Hajslova, J. Advances in high-resolution mass spectrometry based on metabolomics studies for food—A review. *Food Addit. Contam. Part A* **2015**, *32*, 1685–1708. [CrossRef]
27. Papapetros, S.; Louppis, A.; Kosma, I.; Kontakos, S.; Badeka, A.; Kontominas, M.G. Characterization and differentiation of botanical and geographical origin of selected popular sweet cherry cultivars grown in Greece. *J. Food Compos. Anal.* **2018**, *72*, 48–56. [CrossRef]
28. Watanabe, E.; Miyake, S. Direct determination of neonicotinoid insecticides in an analytically challenging crop such as Chinese chives using selective ELISAs. *J. Environ. Sci. Health Part B* **2018**, *53*, 707–712. [CrossRef] [PubMed]
29. Hongsibsong, S.; Prapamontol, T.; Xu, T.; Hammock, B.D.; Wang, H.; Chen, Z.J.; Xu, Z.L. Monitoring of the organophosphate pesticide chlorpyrifos in vegetable samples from local markets in Northern Thailand by developed immunoassay. *Int. J. Environ. Res. Public Health* **2020**, *17*, 4723. [CrossRef] [PubMed]
30. Tan, J.; Li, R.; Jiang, Z.T.; Tang, S.H.; Wang, Y.; Shi, M.; Xiao, Y.Q.; Jia, B.; Lu, T.X.; Wang, H. Synchronous front-face fluorescence spectroscopy for authentication of the adulteration of edible vegetable oil with refined used frying oil. *Food Chem.* **2017**, *217*, 274–280. [CrossRef] [PubMed]
31. Huo, Y.; Kamal, G.M.; Wang, J.; Liu, H.; Zhang, G.; Hu, Z.; Anwar, F.; Du, H. 1H-NMRbased metabolomics for discrimination of rice from different geographical origins of China. *J. Cereal Sci.* **2017**, *76*, 243–252. [CrossRef]
32. Longobardi, F.; Innamorato, V.; Di Gioia, A.; Ventrella, A.; Lippolis, V.; Logrieco, A.F.; Catucci, L.; Agostiano, A. Geographical origin discrimination of lentils (*Lens culinaris* Medik.) using 1H-NMR fingerprinting and multivariate statistical analyses. *Food Chem.* **2017**, *237*, 743–748. [CrossRef]
33. Salazar, M.O.; Pisano, P.L.; Gonzalez Sierra, M.; Furlan, R.L.E. NMR and multivariate data analysis to assess traceability of argentine citrus. *Microchem. J.* **2018**, *141*, 264–270. [CrossRef]
34. Drivelos, S.A.; Georgiou, C.A. Multi-element and multi-isotope-ratio analysis to determine the geographical origin of foods in the European Union. *Trends Anal. Chem.* **2012**, *40*, 38–51. [CrossRef]
35. Mahne Opatić, A.; Nečemer, M.; Lojen, S.; Masten, J.; Zlatić, E.; Šircelj, H.; Stopar, D.; Vidrih, R. Determination of geographical origin of commercial tomato through analysis of stable isotopes, elemental composition and chemical markers. *Food Control* **2018**, *89*, 133–141. [CrossRef]
36. Zambianchi, S.; Soffritti, G.; Stagnati, L.; Patrone, V.; Morelli, L.; Vercesi, A.; Busconi, M. Applicability of DNA traceability along the entire wine production chain in the real case of a large Italian cooperative winery. *Food Control* **2021**, *124*, 107929. [CrossRef]
37. Chedid, E.; Rizou, M.; Kalaitzis, P. Application of high resolution melting combined with DNA-based markers for quantitative analysis of olive oil authenticity and adulteration. *Food Chemistry X* **2020**, *6*, 100082. [CrossRef]
38. Gomes, S.; Breia, R.; Carvalho, T.; Carnide, V.; Martins-Lopes, P. Microsatellite high-resolution melting (SSR-HRM) to track olive genotypes: From field to olive oil. *J. Food Sci.* **2018**, *83*, 2415–2423. [CrossRef] [PubMed]
39. di Rienzo, V.; Fanelli, V.; Miazzi, M.M.; Savino, V.; Pasqualone, A.; Summo, C.; Giannini, P.; Sabetta, W.; Montemurro, C. A reliable analytical procedure to discover table grape DNA adulteration in industrial wines and musts. *Acta Hortic.* **2017**, *1188*, 365–370. [CrossRef]
40. Ben Ayed, R.; Rebai, A. Tunisian table olive oil traceability and quality using SNP genotyping and bioinformatics tools. *BioMed Res. Int.* **2019**, *2019*, 8291341. [CrossRef]
41. Piarulli, L.; Savoia, M.A.; Taranto, F.; D'Agostino, N.; Sardaro, R.; Girone, S.; Gadaleta, S.; Fucili, V.; De Giovanni, C.; Montemurro, C.; et al. A robust DNA isolation protocol from filtered commercial olive oil for PCR-based fingerprinting. *Foods* **2019**, *8*, 462. [CrossRef]
42. Pereira, L.; Gomes, S.; Barrias, S.; Fernandes, J.R.; Martins-Lopes, P. Applying high-resolution melting (HRM) technology to olive oil and wine authenticity. *Food Res. Int.* **2018**, *103*, 170–181. [CrossRef]
43. Boccacci, P.; Chitarra, W.; Schneider, A.; Rolle, L.; Gambino, G. Single-nucleotide polymorphism (SNP) genotyping assays for the varietal authentication of 'Nebbiolo' musts and wines. *Food Chem.* **2020**, *312*, 126100. [CrossRef]
44. Barrias, S.; Fernandes, J.R.; Eiras-Dias, J.E.; Brazão, J.; Martins-Lopes, P. Label free DNA-based optical biosensor as a potential system for wine authenticity. *Food Chem.* **2019**, *270*, 299–304. [CrossRef]
45. Zhang, D.; Vega, F.E.; Infante, F.; Solano, W.; Johnson, E.S.; Meinhardt, L.W. Accurate differentiation of green beans of Arabica and Robusta coffee using nanofluidic array of Single Nucleotide Polymorphism (SNP) markers. *J. AOAC Int.* **2020**, *103*, 315–324. [CrossRef]

46. Silletti, S.; Morello, L.; Gavazzi, F.; Gianì, S.; Braglia, L.; Breviario, D. Untargeted DNA-based methods for the authentication of wheat species and related cereals in food products. *Food Chem.* **2019**, *271*, 410–418. [CrossRef]
47. Morcia, C.; Bergami, R.; Scaramagli, S.; Ghizzoni, R.; Carnevali, P.; Terzi, V. A Chip Digital PCR assay for quantification of common wheat contamination in pasta production chain. *Foods* **2020**, *9*, 911. [CrossRef] [PubMed]
48. Dong, X.; Gao, D.; Dong, J.; Chen, W.; Li, Z.; Wang, J.; Liu, J. Mass ratio quantitative detection for kidney bean in lotus seed paste using duplex droplet digital PCR and chip digital PCR. *Anal. Bioanal. Chem.* **2020**, *412*, 1701–1707. [CrossRef] [PubMed]
49. Swetha, V.P.; Parvathy, V.A.; Sheeja, T.E.; Sasikumar, B. Authentication of *Myristica fragrans* Houtt. using DNA barcoding. *Food Control* **2017**, *73*, 1010–1015. [CrossRef]
50. Uncu, A.O. A trnH-psbA barcode genotyping assay for the detection of common apricot (*Prunus armeniaca* L.) adulteration in almond (*Prunus dulcis* Mill.). *CyTA-J. Food* **2020**, *18*, 187–194. [CrossRef]
51. Lagiotis, G.; Stavridou, E.; Bosmali, I.; Osathanunkul, M.; Haider, N.; Madesis, P. Detection and quantification of cashew in commercial tea products using High Resolution Melting (HRM) analysis. *J. Food Sci.* **2020**, *85*, 1629–1634. [CrossRef] [PubMed]
52. Ding, Y.; Jiang, G.; Huang, L.; Chen, C.; Sun, J.; Zhu, C. DNA barcoding coupled with high-resolution melting analysis for nut species and walnut milk beverage authentication. *J. Sci. Food Agric.* **2020**, *100*, 2372–2379. [CrossRef]
53. Bosmali, I.; Lagiotis, G.; Stavridou, E.; Haider, N.; Osathanunkul, M.; Pasentsis, K.; Madesis, P. Novel authentication approach for coffee beans and the brewed beverage using a nuclear-based species-specific marker coupled with high resolution melting analysis. *LWT-Food Sci. Technol.* **2021**, *137*, 110336. [CrossRef]
54. Valentini, P.; Galimberti, A.; Mezzasalma, V.; De Mattia, F.; Casiraghi, M.; Labra, M.; Pompa, P.P. DNA barcoding meets nanotechnology: Development of a universal colorimetric test for food authentication. *Angew. Chem. Int. Ed.* **2017**, *56*, 8094. [CrossRef] [PubMed]
55. Wu, Y.; Li, M.; Yang, Y.; Jiang, L.; Liu, M.; Wang, B.; Wang, Y. Authentication of small berry fruit in fruit products by DNA barcoding method. *J. Food Sci.* **2018**, *83*, 1494–1504. [CrossRef] [PubMed]
56. Zhao, M.; Wang, B.; Xiang, L.; Xiong, C.; Shi, Y.; Wu, L.; Meng, X.; Dong, G.; Xie, Y.; Sun, W. A novel onsite and visual molecular technique to authenticate saffron (*Crocus sativus*) and its adulterants based on recombinase polymerase amplification. *Food Control* **2019**, *100*, 117–121. [CrossRef]
57. Zhao, M.; Shi, Y.; Wu, L.; Guo, L.; Liu, W.; Xiong, C.; Yan, S.; Sun, W.; Chen, S. Rapid authentication of the precious herb saffron by loop-mediated isothermal amplification (LAMP) based on internal transcribed spacer 2 (ITS2) sequence. *Sci. Rep.* **2016**, *6*, 25370. [CrossRef]
58. Cibecchini, G.; Cecere, P.; Tumino, G.; Morcia, C.; Ghizzoni, R.; Carnevali, P.; Terzi, V.; Pompa, P.P. A fast, naked-eye assay for varietal traceability in the durum wheat production chain. *Foods* **2020**, *9*, 1691. [CrossRef] [PubMed]
59. Hu, Y.; Lu, X. Rapid pomegranate juice authentication using a simple sample-to-answer hybrid paper/polymer-based lab-on-a-chip device. *ACS Sens.* **2020**, *5*, 2168–2176. [CrossRef]
60. Lo, Y.T.; Shaw, P.C. Application of next-generation sequencing for the identification of herbal products. *Biotechnol. Adv.* **2019**, *37*, 107450. [CrossRef]
61. Raime, K.; Krjutškov, K.; Remm, M. Method for the identification of plant DNA in food using alignment-free analysis of sequencing reads: A case study on lupin. *Front. Plant Sci.* **2020**, *11*, 646. [CrossRef]
62. Lee, H.J.; Koo, H.J.; Lee, J.; Lee, S.C.; Lee, D.Y.; Giang, V.N.L.; Kim, M.; Shim, H.; Park, J.Y. Yoo, K.O.; et al. Authentication of *Zanthoxylum* species based on integrated analysis of complete chloroplast genome sequences and metabolite profiles. *J. Agric. Food Chem.* **2017**, *65*, 10350–10359. [CrossRef]
63. Kim, Y.; Shin, J.; Oh, D.R.; Kim, D.W.; Lee, H.S.; Choi, C. Complete chloroplast genome sequences of *Vaccinium bracteatum* Thunb., *V. vitis-idaea* L., and *V. uliginosum* L. (Ericaceae). *Mitochondrial DNA B* **2020**, *5*, 1843–1844. [CrossRef]
64. Khansaritoreh, E.; Salmaki, Y.; Ramezani, E.; Akbari Azirani, T.; Keller, A.; Neumann, K.; Alizadeh, K.; Zarre, S.; Beckh, G.; Behling, H. Employing DNA metabarcoding to determine the geographical origin of honey. *Heliyon* **2020**, *6*, e05596. [CrossRef] [PubMed]
65. Beltramo, C.; Cerutti, F.; Brusa, F.; Mogliotti, P.; Garrone, A.; Squadrone, S.; Acutis, P.L.; Peletto, S. Exploring the botanical composition of polyfloral and monofloral honeys through DNA metabarcoding. *Food Control* **2021**, *128*, 108175.
66. Powell, W.; Machray, G.C.; Provan, J. Polymorphism revealed by simple sequence repeats. *Trends Plant Sci.* **1996**, *1*, 215–222. [CrossRef]
67. Stagnati, L.; Soffritti, G.; Martino, M.; Bortolini, C.; Lanubile, A.; Busconi, M.; Marocco, A. Cocoa beans and liquor fingerprinting: A real case involving SSR profiling of CCN51 and "Nacional" varieties. *Food Control* **2020**, *118*, 107392. [CrossRef]
68. Pinczinger, D.; von Reth, M.; Hanke, M.V.; Flachowsky, H. SSR fingerprinting of raspberry cultivars traded in Germany clearly showed that certainty about the genotype authenticity is a prerequisite for any horticultural experiment. *Eur. J. Hortic. Sci.* **2020**, *85*, 79–85. [CrossRef]
69. Sabetta, W.; Miazzi, M.M.; di Rienzo, V.; Fanelli, V.; Pasqualone, A.; Montemurro, C. Development and application of protocols to certify the authenticity and traceability of Apulian typical products in olive sector. *Riv. Ital. Delle Sostanze Grasse* **2017**, *94*, 37–43.
70. Verdone, M.; Rao, R.; Coppola, M.; Corrado, G. Identification of zucchini varieties in commercial food products by DNA typing. *Food Control* **2018**, *84*, 197–204. [CrossRef]
71. Ganopoulos, I.; Argiriou, A.; Tsaftaris, A. Microsatellite high resolution melting (SSR-HRM) analysis for authenticity testing of protected designation of origin (PDO) sweet cherry products. *Food Control* **2011**, *22*, 532–541. [CrossRef]

72. Bosmali, I.; Ganopoulos, I.; Madesis, P.; Tsaftaris, A. Microsatellite and DNA-barcode regions typing combined with High Resolution Melting (HRM) analysis for food forensic uses: A case study on lentils (*Lens culinaris*). *Food Res. Int.* **2012**, *46*, 141–147. [CrossRef]
73. di Rienzo, V.; Miazzi, M.M.; Fanelli, V.; Savino, V.; Pollastro, S.; Colucci, F.; Miccolupo, A.; Blanco, A.; Pasqualone, A.; Montemurro, C. An enhanced analytical procedure to discover table grape DNA adulteration in industrial musts. *Food Control* **2016**, *60*, 124–130. [CrossRef]
74. Pasqualone, A.; Montemurro, C.; Grinn-Gofron, A.; Sonnante, G.; Blanco, A. Detection of soft wheat in semolina and durum wheat bread by analysis of DNA microsatellites. *J. Agric. Food Chem.* **2007**, *55*, 3312–3318. [CrossRef] [PubMed]
75. Zhao, J.; Li, A.; Jin, X.; Pan, L. Technologies in individual animal identification and meat products traceability. *Biotechnol. Biotechnol. Equip.* **2020**, *34*, 48–57. [CrossRef]
76. Marrano, A.; Birolo, G.; Prazzoli, M.L.; Lorenzi, S.; Valle, G.; Grando, M.S. SNP-discovery by RAD-sequencing in a germplasm collection of wild and cultivated grapevines (*V. vinifera* L.). *PLoS ONE* **2017**, *12*, e0170655. [CrossRef] [PubMed]
77. Taranto, F.; D'Agostino, N.; Pavan, S.; Fanelli, V.; di Rienzo, V.; Sabetta, W.; Miazzi, M.M.; Zelasco, S.; Perri, E.; Montemurro, C. Single nucleotide polymorphism (SNP) diversity in an olive germplasm collection. *Acta Hortic.* **2018**, *1199*, 27–32. [CrossRef]
78. Pavan, S.; Bardaro, N.; Fanelli, V.; Marcotrigiano, A.R.; Mangini, G.; Taranto, F.; Catalano, D.; Montemurro, C.; De Giovanni, C.; Lotti, C.; et al. Genotyping by Sequencing of cultivated lentil (*Lens culinaris* Medik.) highlights population structure in the Mediterranean gene pool associated with geographic patterns and phenotypic variables. *Front. Genet.* **2019**, *10*, 872. [CrossRef]
79. Singh, R.; Iquebal, M.A.; Mishra, C.N.; Jaiswal, S.; Kumar, D.; Raghav, N.; Paul, S.; Sheoran, S.; Sharma, P.; Gupta, A.; et al. Development of model web-server for crop variety identification using throughput SNP genotyping data. *Sci. Rep.* **2019**, *9*, 5122. [CrossRef]
80. Akpertey, A.; Padi, F.K.; Meinhardt, L.; Zhang, D. Effectiveness of Single Nucleotide Polymorphism markers in genotyping germplasm collections of Coffea canephora using KASP assay. *Front. Plant Sci.* **2021**, *11*, 612593. [CrossRef]
81. Spaniolas, S.; Bazakos, C.; Tucker, G.A.; Bennett, M.J. Comparison of SNP-based detection assays for food analysis: Coffee authentication. *J. AOAC Int.* **2014**, *97*, 4. [CrossRef]
82. Fang, W.; Meinhardt, L.W.; Mischke, S.; Bellato, C.M.; Motilal, L.; Zhang, D. Accurate determination of genetic identity for a single cacao bean, using molecular markers with a nanofluidic system, ensures cocoa authentication. *J. Agric. Food Chem.* **2014**, *62*, 481–487. [CrossRef]
83. Fang, W.P.; Meinhardt, L.W.; Tan, H.W.; Zhou, L.; Mischke, S.; Zhang, D. Varietal identification of tea (*Camellia sinensis*) using nanofluidic array of single nucleotide polymorphism (SNP) markers. *Hortic. Res.* **2014**, *1*, 14035. [CrossRef]
84. Nielsen, R.; Paul, J.S.; Albrechtsen, A.; Song, Y.S. Genotype and SNP calling from next-generation sequencing data. *Nat. Rev. Genet.* **2011**, *12*, 443–451. [CrossRef]
85. Galimberti, A.; De Mattia, F.; Losa, A.; Bruni, I.; Federici, S.; Casiraghi, M.; Martellos, S.; Labra, M. DNA barcoding as a new tool for food traceability. *Food Res. Int.* **2013**, *50*, 55–63. [CrossRef]
86. Sonnante, G.; Montemurro, C.; Morgese, A.; Sabetta, W.; Blanco, A.; Pasqualone, A. DNA microsatellite region for a reliable quantification of ~soft wheat adulteration in durum wheat-based foodstuffs by real-time PCR. *J. Agric. Food Chem.* **2009**, *57*, 10199–1020411. [CrossRef] [PubMed]
87. Matsuoka, Y.; Arami, S.; Sato, M.; Haraguchi, H.; Kurimoto, Y.; Imai, S.; Tanaka, K.; Mano, J.; Furui, S.; Kitta, K. Development of methods to distinguish between durum/common wheat and common wheat in blended flour using PCR. *J. Food Hyg. Soc. Jpn.* **2012**, *53*, 195–202. [CrossRef] [PubMed]
88. Voorhuijzen, M.M.; van Dijk, J.P.; Prins, T.W.; Van Hoef, A.M.A.; Seyfarth, R.; Kok, E.J. Development of a multiplex DNA-based traceability tool for crop plant materials. *Anal. Bioanal. Chem.* **2012**, *402*, 693–701. [CrossRef] [PubMed]
89. Morisset, D.; Stebih, D.; Milavec, M.; Gruden, K.; Zel, J. Quantitative analysis of food and feed samples with droplet digital PCR. *PLoS ONE* **2013**, *8*, e62583. [CrossRef]
90. Low, H.; Chan, S.J.; Soo, G.H.; Ling, B.; Tan, E.L. Clarity™ digital PCR system: A novel platform for absolute quantification of nucleic acids. *Anal. Bioanal. Chem.* **2017**, *409*, 1869–1875. [CrossRef] [PubMed]
91. Demeke, T.; Dobnik, D. Critical assessment of digital PCR for the detection and quantification of genetically modified organisms. *Anal. Bioanal. Chem.* **2018**, *410*, 4039–4050. [CrossRef]
92. He, L.; Simpson, D.J.; Gänzle, M.G. Detection of enterohaemorrhagic Escherichia coli in food by droplet digital PCR to detect simultaneous virulence factors in a single genome. *Food Microbiol.* **2020**, *90*, 103466. [CrossRef]
93. Pierboni, E.; Rondini, C.; Torricelli, M.; Ciccone, L.; Tovo, G.R.; Mercuri, M.L.; Altissimi, S.; Haouet, N. Digital PCR for analysis of peanut and soybean allergens in foods. *Food Control* **2018**, *92*, 128–136. [CrossRef]
94. Hebert, P.D.N.; Ratnasingham, S.; deWaard, J.R. Barcoding animal life: Cytochrome c oxidase subunit 1 divergences among closely related species. *Proc. Royal Soc. B-Biol. Sci.* **2003**, *270*, S96–S99. [CrossRef]
95. Liu, Y.; Wang, X.; Wang, L.; Chen, X.; Pang, X.; Han, J. A nucleotide signature for the identification of American ginseng and its products. *Front. Plant Sci.* **2016**, *18*, 319. [CrossRef] [PubMed]
96. Uncu, A.T.; Uncu, A.O. Plastid trnH-psbA intergenic spacer serves as a PCR-based marker to detect common grain adulterants of coffee (*Coffea arabica* L.). *Food Control* **2018**, *91*, 32–39. [CrossRef]
97. Bosmali, I.; Ordoudi, S.A.; Tsimidou, M.Z.; Madesis, P. Greek PDO saffron authentication studies using species specific molecular markers. *Food Res. Int.* **2017**, *100*, 899–907. [CrossRef] [PubMed]

98. Liu, Y.; Xiang, L.; Zhang, Y.; Lai, X.; Xiong, C.; Li, J.; Su, Y.; Sun, W.; Chen, S. DNA barcoding based identification of *Hippophae* species and authentication of commercial products by high resolution melting analysis. *Food Chem.* **2018**, *242*, 62–67. [CrossRef] [PubMed]
99. Ballin, N.Z.; Onaindia, J.O.; Jawad, H.; Fernandez-Carazo, R.; Maquet, A. High-resolution melting of multiple barcode amplicons for plant species authentication. *Food Control* **2019**, *105*, 141–150. [CrossRef]
100. Gong, L.; Qiu, X.H.; Huang, J.; Xu, W.; Bai, J.Q.; Zhang, J.; Su, H.; Xu, C.M.; Huang, Z.H. Constructing a DNA barcode reference library for southern herbs in China: A resource for authentication of southern Chinese medicine. *PLoS ONE* **2018**, *13*, e0201240.
101. Kane, N.C.; Cronk, Q. Botany without borders, barcoding in focus. *Mol. Ecol.* **2008**, *17*, 5175–5176. [CrossRef]
102. Kane, N.; Sveinsson, S.; Dempewolf, H.; Yang, J.Y.; Zhang, D.; Engels, J.M.M.; Cronk, Q. Ultra-barcoding in cacao (*Theobroma* spp.; *Malvaceae*) using whole chloroplast genomes and nuclear ribosomal DNA. *Am. J. Bot.* **2012**, *99*, 320–329. [CrossRef] [PubMed]
103. Ratnasingham, S.; Hebert, P.D.N. BOLD: The Barcode of Life Data System (http://www.barcodinglife.org). *Mol. Ecol. Notes* **2007**, *7*, 355–364. [CrossRef]
104. Leonardo, S.; Toldrà, A.; Campàs, M. Biosensors based on isothermal DNA amplification for bacterial detection in food safety and environmental monitoring. *Sensors* **2021**, *21*, 602. [CrossRef]
105. Santiago-Felipe, S.; Tortajada-Genaro, L.A.; Puchades, R.; Maquieira, A. Recombinase polymerase and enzyme-linked immunosorbent assay as a DNA amplification-detection strategy for food analysis. *Anal. Chim. Acta* **2014**, *811*, 81–87. [CrossRef] [PubMed]
106. Zhang, X.; Lowe, S.B.; Gooding, J.J. Brief review of monitoring methods for loop-mediated isothermal amplification (LAMP). *Biosens. Bioelectron.* **2014**, *61*, 491–499. [CrossRef] [PubMed]
107. Wong, Y.P.; Othman, S.; Lau, Y.L.; Radu, S.; Chee, H.Y. Loop-mediated isothermal amplification (LAMP): A versatile technique for detection of micro-organisms. *J. Appl. Microbiol.* **2017**, *124*, 626–643. [CrossRef] [PubMed]
108. Li, R.; Wang, C.; Ji, L.; Zhao, X.X.; Liu, M.; Zhang, D.; Shi, J. Loop mediated isothermal amplification (LAMP) assay for GMO detection: Recent progresses and future perspectives. *Open Access Libr. J.* **2015**, *2*, e1264.
109. Wilkinson, M.J.; Szabo, C.; Ford, C.S.; Yarom, Y.; Croxford, A.E.; Camp, A.; Gooding, P. Replacing Sanger with Next Generation Sequencing to improve coverage and quality of reference DNA barcodes for plants. *Sci. Rep.* **2017**, *7*, 46040. [CrossRef]
110. Kumar, K.R.; Cowley, M.J.; Davis, R.L. Next-Generation Sequencing and emerging technologies. *Semin. Thromb. Hemost.* **2019**, *45*, 661–673. [CrossRef]
111. Ripp, F.; Krombholz, C.F.; Liu, Y.; Weber, M.; Schäfer, A.; Schmidt, B.; Köppel, R.; Hankeln, T. All-Food-Seq (AFS): A quantifiable screen for species in biological samples by deep DNA sequencing. *BMC Genom.* **2014**, *15*, 639. [CrossRef]
112. Beck, K.L.; Haiminen, N.; Chambliss, D.; Edlund, S.; Kunitomi, M.; Huang, B.C.; Kong, N.; Ganesan, B.; Baker, R.; Markwell, P.; et al. Monitoring the microbiome for food safety and quality using deep shotgun sequencing. *NPJ Sci Food* **2021**, *5*, 3. [CrossRef]
113. Haynes, E.; Jimenez, E.; Pardo, M.A.; Helyar, S.J. The future of NGS (Next Generation Sequencing) analysis in testing food authenticity. *Food Control* **2019**, *101*, 134–143. [CrossRef]
114. Haiminen, N.; Edlund, S.; Chambliss, D.; Kunitomi, M.; Weimer, B.C.; Ganesan, B.; Baker, R.; Markwell, P.; Davis, M.; Huang, C.; et al. Food authentication from shotgun sequencing reads with an application on high protein powders. *NPJ Sci Food* **2019**, *3*, 24. [CrossRef]
115. Zhang, N.; Ramachandran, P.; Wen, J.; Duke, J.A.; Metzman, H.; McLaughlin, W.; Ottesen, A.R.; Timme, R.E.; Handy, S.M. Development of a reference standard library of chloroplast genome sequences, GenomeTrakrCP. *Planta Med.* **2017**, *83*, 1420–1430. [CrossRef]
116. Bruno, A.; Sandionigi, A.; Agostinetto, G.; Bernabovi, L.; Frigerio, J.; Casiraghi, M.; Labra, M. Food tracking perspective: DNA metabarcoding to identify plant composition in complex and processed food products. *Genes* **2019**, *10*, 248. [CrossRef]
117. Speranskaya, A.S.; Khafizov, K.; Ayginin, A.A.; Krinitsina, A.A.; Omelchenko, D.O.; Nilova, M.V.; Severova, E.E.; Samokhina, E.N.; Shipulin, G.A.; Logacheva, M.D. Comparative analysis of Illumina and Ion Torrent high-throughput sequencing platforms for identification of plant components in herbal teas. *Food Control* **2018**, *93*, 315–324. [CrossRef]
118. Frigerio, J.; Gorini, T.; Galimberti, A.; Bruni, I.; Tommasi, N.; Mezzasalma, V.; Labra, M. DNA barcoding to trace Medicinal and Aromatic Plants from the field to the food supplement. *J. Appl. Bot. Food Qual.* **2019**, *92*, 33–38.
119. Banchi, E.; Ametrano, C.G.; Greco, S.; Stanković, D.; Muggia, L.; Pallavicini, A. PLANiTS: A curated sequence reference dataset for plant ITS DNA metabarcoding. *Database* **2020**, *2020*, baz155. [CrossRef] [PubMed]
120. Utzeri, V.J.; Ribani, A.; Schiavo, G.; Bertolini, F.; Bovo, S.; Fontanesi, L. Application of next generation semiconductor based sequencing to detect the botanical composition of monofloral, polyfloral and honeydew honey. *Food Control* **2018**, *86*, 342–349. [CrossRef]
121. Gostel, M.R.; Zúñiga, J.D.; Kress, W.J.; Funk, V.A.; Puente-Lelievre, C. Microfluidic Enrichment Barcoding (MEBarcoding): A new method for high throughput plant DNA barcoding. *Sci. Rep.* **2020**, *10*, 8701. [CrossRef] [PubMed]
122. Xia, Y.; Chen, F.; Jiang, L.; Li, S.; Zhang, J. Development of an efficient method to extract DNA from refined soybean oil. *Food Anal. Methods* **2021**, *14*, 196–207. [CrossRef]
123. Pereira, L.; Guedes-Pinto, H.; Martins-Lopes, P. An enhanced method for *Vitis vinifera* L. DNA extraction from wines. *Am. J. Enol. Vitic.* **2011**, *62*, 4. [CrossRef]
124. Bojang, K.P.; Kuna, A.; Pushpavalli, S.N.C.V.L.; Sarkar, S.; Sreedhar, M. Evaluation of DNA extraction methods for molecular traceability in cold pressed, solvent extracted and refined groundnut oils. *J. Food Sci. Technol.* **2021**, 1–7. [CrossRef]

125. Soares, S.; Amaral, J.S.; Oliveira, M.B.P.P.; Mafra, I. Improving DNA isolation from honey for the botanical origin identification. *Food Control* **2015**, *48*, 130–136. [CrossRef]
126. López-Calleja, I.M.; de la Cruz, S.; Martín, R.; González, I.; García, T. Duplex real-time PCR method for the detection of sesame (*Sesamum indicum*) and flaxseed (*Linum usitatissimum*) DNA in processed food products. *Food Addit. Contam. Part A* **2015**, *32*, 1772–1785. [CrossRef] [PubMed]
127. Kumar, P.; Gupta, V.; Misra, A.; Modi, D.; Pandey, B. Potential of molecular markers in plant biotechnology. *Plant Omics* **2009**, *4*, 141–162.
128. Jagadeesan, B.; Gerner-Smidt, P.; Allard, M.W.; Leuillet, S.; Winkler, A.; Xiao, Y.; Chaffron, S.; Van Der Vossen, J.; Tang, S.; Katase, M.; et al. The use of next generation sequencing for improving food safety: Translation into practice. *Food Microbiol.* **2019**, *79*, 96–115. [CrossRef]
129. Galimberti, A.; Bruno, A.; Mezzasalma, V.; De Mattia, F.; Bruni, I.; Labra, M. Emerging DNA-based technologies to characterize food ecosystems. *Food Res. Int.* **2015**, *69*, 424–433. [CrossRef]
130. Besse, P.; Da Silva, D.; Grisoni, M. Plant DNA barcoding principles and limits: A case study in the Genus Vanilla. In *Molecular Plant Taxonomy*; Besse, P., Ed.; Springer: New York, NY, USA, 2021; Volume 2222, pp. 131–148.

Commentary

Digested Civet Coffee Beans (Kopi Luwak)—An Unfortunate Trend in Specialty Coffee Caused by Mislabeling of *Coffea liberica*?

Dirk W. Lachenmeier [1],* and Steffen Schwarz [2]

1. Chemisches und Veterinäruntersuchungsamt (CVUA) Karlsruhe, Weissenburger Strasse 3, 76187 Karlsruhe, Germany
2. Coffee Consulate, Hans-Thoma-Strasse 20, 68163 Mannheim, Germany; schwarz@coffee-consulate.com
* Correspondence: lachenmeier@web.de; Tel.: +49-721-926-5434

Abstract: In the context of animal protection, the trend of digested coffees such as Kopi Luwak produced by civet cats in captivity should not be endorsed. Previous studies on such coffees may have been flawed by sample selection and misclassification. As wild civets may prefer *Coffea liberica* beans, due to their higher sugar content, the chemical differences may be caused by the *Coffea* species difference combined with a careful selection of ripe, defect-free cherries by the animals, rather than changes caused by digestion. This may also explain the observed differences between Kopi Luwak from wild civets (mainly *C. liberica*) compared to the one from animals in captivity (typically fed with *C. arabica* and/or *C. canephora*).

Keywords: coffee; fermentation; gastrointestinal tract; Kopi Luwak; civet

1. Introduction

The topic of digested coffees is currently receiving a renewed interest and has recently been proposed as a "new trend in specialty coffee" [1].

In this commentary, we want to point out the interesting issue of *Coffea* species assignment in the context of digested coffee studies. The first problem emerges when chemical studies are conducted in non-coffee growing countries, and sampling relies on commercial suppliers, often with doubtful authenticity. The control group is also problematic in the digested coffee studies, as wild civets may select the sweetest, most ripe, and healthy cherries, while the control coffee of commercial quality may include different stages of ripeness and the typical amount of defective beans. For example, it makes no sense to use a Brazilian *C. arabica* coffee as control group for Kopi Luwak from Indonesia. Geographical and variety differences within *C. arabica* alone may explain the observed differences.

The second problem with digested coffees is that many studies may have missed that the actual coffee species under investigation has been *Coffea liberica*, not *Coffea arabica* or *Coffea canephora*, which has been incorrectly assumed. This hypothesis was first raised during an international roasting competition for Liberica coffee [2].

2. A Short Critique of Previous Digested Coffee Studies

The study of Marcone [3] is currently the most widely cited study on digested coffees according to Google Scholar (187 citations in June 2021). Marcone [3] obtained Kopi Luwak and control beans (not having gone through the palm civet) from a supplier in California. Both the Kopi Luwak and control coffee beans were claimed as being Indonesian *Coffea canephora* var. robusta. The study also included African civet coffee collected in western Ethiopia. No species was provided for the Ethiopian coffee, which, however, should be assumed as being *Coffea arabica*, the predominant species in Ethiopia. Marcone [3] provided photographs of the studied beans (reproduced in Figure 1a–c).

Figure 1. Photographs of coffees claimed as being digested: (**a**) Kopi Luwak coffee beans (claimed as being *Coffea canephora* var. robusta), (**b**) Nekemte-African Civet coffee beans, and (**c**) Abdela-African Civet coffee beans. Photographs of non-digested coffees for comparison: (**d**) *Coffea liberica*, (**e**) *Coffea canephora* var. Old Paradenia (India), and (**f**) *Coffea arabica* var. Catuaí Vermelho (Brasil). ((**a**–**c**) reprinted with graphical improvement (background and noise removed) from Food Research International, 37, Massimo F. Marcone, Composition and properties of Indonesian palm civet coffee (Kopi Luwak) and Ethiopian civet coffee, pp. 901–912 [3], Copyright (2004), with permission from Elsevier. (**d**–**f**) are original photographs).

According to Marcone [3], the beans were assigned as *C. canephora* (Figure 1a), and two types of Ethiopian coffee (Figure 1b,c). However according to our assessment of the shapes, the beans are actually *C. liberica* (Figure 1a), *C. canephora* (which is rather unusual for Ethiopia, therefore assumed as an adulterated product) (Figure 1b), and *C. arabica* (Figure 1c). Please note the bulging and raised nature of the beans at the cut for liberica (Figure 1a). Arabica and canephora are flat at the cut and equally high on both sides. In our opinion, the mislabeling is quite clear. For comparison purposes, we provide examples of authentic *C. liberica* (Figure 1d), *C. canephora* (Figure 1e) and *C. arabica* (Figure 1f). The fact that *C. liberica* exhibits such a little-noticed existence is surely one of the reasons why this circumstance escaped the authors of Kopi Luwak studies and reviews [1] thus far. The species difference may also explain the different surface morphology of the beans [3]. The discrimination ability of some analytical methods can also be explained in that two different coffee species were compared against each other (i.e., *C. liberica* in Kopi Luwak vs. *C. arabica* as control group, e.g., compare Jumhawan et al. [4–6] and Suhandy and Yulia [7]).

3. Kopi Luwak a *Coffea liberica* in Disguise?

The distinctly different taste and highly valued flavor of Kopi Luwak coffee may be caused by the pure fact that it is *Coffea liberica*, which has a completely different flavor, with very complex profile compared to the commercial coffee species *C. arabica* and *C. canephora*. *C. liberica* has the highest sugar content of all coffees, and thus has the highest risk of fermentation. The sugar content may also be the reason that the civets and other coffee consuming animals prefer *C. liberica* over the other species, if they are available in the same area.

Diligently prepared *C. liberica* shows intense fruity and floral notes (strawberry, jackfruit, mango, banana) and a lactic character (yogurt, cream, mascarpone, crème fraiche) with a pronounced body and intense sweetness. When roasted too dark, the coffee offers notes that reach into the realm of ripe, sweet blue cheese and cheddar.

The lactic, cheesy, perhaps also animalic character of *C. liberica* may be easily misinterpreted as an influence potentially caused by animal digestion or intra-animal fermentation (i.e., the alleged change in taste caused by digestive enzymes of the animals), which are not convincingly proven in previous scientific studies. Currently, there are no sensory or chemical studies available investigating the possibility to distinguish Kopi Luwak from regular coffee prepared from *Coffea liberica* species.

One of the first descriptions of Kopi Luwak, from Brehm in 1883 [8], suggested that the civet released the undigested seeds, that the excrement consisted entirely of caked, but incidentally undamaged coffee beans, and that the animals provide the very best coffee because they ate the ripest fruits. This description stands largely unchallenged to this day, and the scientific proof for the alternative hypothesis, that animal digestion actually changes the coffee and its flavor profile, so far lacks convincing proof. Due to the animal cruelty involved, we believe that this question does not necessarily need further investigation. Ripe and sweet coffee cherries of *C. liberica* may be selected by means other than the use of animals.

4. Conclusions

The authors believe that digested coffee is rather a perverted trend in specialty coffee, especially if the civet cats are kept in captivity purely for the purpose of coffee production [9] (Figure 2). In this regard, it is almost a relief that much coffee labelled as "Kopi Luwak" is probably a counterfeited product that has never seen the digestive tract of an animal (42% of Kopi Luwak were claimed as being found to be either complete fakes or adulterated with regular coffee beans [10]).

Figure 2. Civet kept caged for Kopi Luwak production (attribution: author Surtr, (https://commons.wikimedia.org/wiki/File:Luwak_(civet_cat)_in_cage.jpg accessed on 8 June 2021) license CC BY-SA 2.0, (https://creativecommons.org/licenses/by-sa/2.0/ accessed on 8 June 2021) via Wikimedia Commons).

Hopefully, the observation that an already valued specialty coffee such as Kopi Luwak may actually be *Coffea liberica* will encourage a new debate on this species in coffee cultivation, especially against the backdrop of climate change. It would certainly be desirable for the diversity of flavors in coffee, as well as avoid animal cruelty for an unnecessary procedure.

Author Contributions: Conceptualization, D.W.L. and S.S.; writing—original draft preparation, D.W.L.; writing—review and editing, S.S. All authors have read and agreed to the published version of the manuscript.

Funding: This research received no external funding.

Data Availability Statement: No new data were created or analyzed in this study. Data sharing is not applicable to this article.

Conflicts of Interest: S.S. is owner of Coffee Consulate, Mannheim, Germany. Coffee Consulate is an independent training and research center. Coffee Consulate is not commercializing digested coffees. Therefore, S.S. reports no conflicts of interest related to the work under consideration. D.W.L. declares no conflicts of interest.

References

1. Raveendran, A.; Murthy, P.S. New trends in specialty coffees—"The digested coffees". *Crit. Rev. Food Sci. Nutr.* **2021**, *61*. [CrossRef]
2. Coffee Consulate. *Liberica Coffee International Roasting Competition 2019*; Coffee Consulate: Mannheim, Germany, 2019; Available online: https://www.coffee-consulate.com/media/09/59/e8/1592404961/Booklet_LibericaRoastingCompetition2019.pdf (accessed on 8 June 2021).
3. Marcone, M.F. Composition and properties of Indonesian palm civet coffee (Kopi Luwak) and Ethiopian civet coffee. *Food Res. Int.* **2004**, *37*, 901–912. [CrossRef]
4. Jumhawan, U.; Putri, S.P.; Yusianto; Bamba, T.; Fukusaki, E. Quantification of coffee blends for authentication of Asian palm civet coffee (Kopi Luwak) via metabolomics: A proof of concept. *J. Biosci. Bioeng.* **2016**, *122*, 79–84. [CrossRef] [PubMed]
5. Jumhawan, U.; Putri, S.P.; Yusianto; Bamba, T.; Fukusaki, E. Application of gas chromatography/flame ionization detector-based metabolite fingerprinting for authentication of Asian palm civet coffee (Kopi Luwak). *J. Biosci. Bioeng.* **2015**, *120*, 555–561. [CrossRef] [PubMed]
6. Jumhawan, U.; Putri, S.P.; Yusianto; Marwani, E.; Bamba, T.; Fukusaki, E. Selection of discriminant markers for authentication of Asian palm civet coffee (Kopi Luwak): A metabolomics approach. *J. Agric. Food Chem.* **2013**, *61*, 7994–8001. [CrossRef] [PubMed]
7. Suhandy, D.; Yulia, M. The use of partial least square regression and spectral data in UV-visible region for quantification of adulteration in Indonesian palm civet coffee. *Int. J. Food Sci.* **2017**, *2017*, 6274178. [CrossRef] [PubMed]
8. Brehm, A. Musang (*Paradoxurus fasciatus*). In *Brehms Thierleben. Allgemeine Kunde des Thierreichs, Zweiter Band, Erste Abtheilung: Säugethiere, Dritter Band: Hufthiere, Seesäugethiere*; Verlag des Bibliographischen Instituts: Leipzig, Germany, 1883; pp. 31–34. Available online: http://www.zeno.org/nid/20007931662 (accessed on 8 June 2021). (In German)
9. Lynn, G.; Rogers, C. *Civet Cat Coffee's Animal Cruelty Secrets*; BBC News: London, UK, 2013; Available online: https://www.bbc.com/news/uk-england-london-24034029 (accessed on 8 June 2021).
10. Davies, D. *Eat Up—And Don't Forget the Palate Cleansers*; NPR: Washington, DC, USA, 2007; Available online: https://www.npr.org/templates/story/story.php?storyId=11847227 (accessed on 8 June 2021).

Communication

Potential of Ultraviolet-Visible Spectroscopy for the Differentiation of Spanish Vinegars According to the Geographical Origin and the Prediction of Their Functional Properties

Raúl González-Domínguez [1,2,*], Ana Sayago [1,2] and Ángeles Fernández-Recamales [1,2]

[1] AgriFood Laboratory, Faculty of Experimental Sciences, University of Huelva, 21007 Huelva, Spain; ana.sayago@dqcm.uhu.es (A.S.); recamale@dqcm.uhu.es (Á.F.-R.)
[2] International Campus of Excellence CeiA3, University of Huelva, 21007 Huelva, Spain
* Correspondence: raul.gonzalez@dqcm.uhu.es; Tel.: +34-959219975

Citation: González-Domínguez, R.; Sayago, A.; Fernández-Recamales, Á. Potential of Ultraviolet-Visible Spectroscopy for the Differentiation of Spanish Vinegars According to the Geographical Origin and the Prediction of Their Functional Properties. *Foods* **2021**, *10*, 1830. https://doi.org/10.3390/foods10081830

Academic Editors: Christopher John Smith and Simon Haughey

Received: 18 June 2021
Accepted: 5 August 2021
Published: 7 August 2021

Publisher's Note: MDPI stays neutral with regard to jurisdictional claims in published maps and institutional affiliations.

Copyright: © 2021 by the authors. Licensee MDPI, Basel, Switzerland. This article is an open access article distributed under the terms and conditions of the Creative Commons Attribution (CC BY) license (https://creativecommons.org/licenses/by/4.0/).

Abstract: High-quality wine vinegars with unique organoleptic characteristics are produced in southern Spain under three Protected Designations of Origin (PDO), namely "Jerez", "Condado de Huelva" and "Montilla-Moriles". To guarantee their authenticity and avoid frauds, robust and low-cost analytical methodologies are needed for the quality control and traceability of vinegars. In this study, we propose the use of ultraviolet-visible spectroscopy in combination with multivariate statistical tools to discriminate Spanish wine vinegars according to their geographical origin, as well as to predict their physicochemical and functional properties. Linear discriminant analysis provided a clear clustering of vinegar samples according to the PDO with excellent classification performance (98.6%). Furthermore, partial least squares regression analysis demonstrated that spectral data can serve as accurate predictors of the total phenolic content and antioxidant activity of vinegars. Accordingly, UV-Vis spectroscopy stands out as a suitable analytical tool for simple and rapid authentication and traceability of vinegars.

Keywords: vinegar; protected designation of origin; UV-Vis spectroscopy; authentication; prediction

1. Introduction

Vinegar is a condiment widely employed in the Mediterranean and Asian diets to preserve and improve the sensory characteristics of foods. According to the Codex Alimentarius Commission, vinegars can be obtained from different agricultural products rich in starch and/or sugars by means of a double fermentation process (i.e., alcoholic and acetic fermentation). In Europe, grapes, apple, pomegranate, cherry and other fruit juices are the most commonly used raw materials for vinegar production, whereas Asian vinegars are normally based on cereals such as sorghum, rice, sticky rice and others [1]. Spain is one of the major producers of high-quality wine vinegars worldwide, whose production is mainly concentrated in Andalusia (Southern Spain). A singularity of the wines and vinegars produced in this geographical area is the use of the traditional aging system of "criaderas and soleras" [2], which provides them with unique organoleptic characteristics that are highly appreciated by consumers. As a result of these unique characteristics, three Protected Designations of Origin (PDO) of Andalusian vinegars have been recognized in accordance with the European Community legislation (Council Regulation (EC) No 510/2006) [3], namely the PDO "Vinagre de Jerez", registered in 1995; the PDO "Vinagre del Condado de Huelva", registered in 2002; and the PDO "Vinagre de Montilla-Moriles", registered in 2008 [4–6].

The composition and organoleptic characteristics of vinegars are influenced by multiple factors, such as the raw material used as a substrate, the acetification system, or the time and method employed for vinegar aging in wooden barrels (e.g., "criaderas and soleras" system, "añada" system), among others [7,8]. These chemical variations, mainly in terms of organic acids and polyphenolic compounds, can in turn affect the functional properties of vinegars (e.g., acidity, antioxidant activity). In this respect, Budak et al. have reported that functional and therapeutic properties of vinegar on human health comprise antibacterial activity, blood pressure reduction, antioxidant activity, prevention of cardiovascular diseases, and improved blood glucose response [9]. Altogether, it becomes evident that the food industry requires robust analytical methodologies to characterize the quality and verify the geographical origin of vinegars. In this context, several studies have previously reported the possibility of discriminating vinegars by using classical targeted approaches based on atomic emission spectroscopy, gas chromatography–mass spectrometry and high-performance liquid chromatography for the determination of mineral elements [10], volatile compounds [11–13], as well as polyphenols, organic acids and amino acids [13,14], respectively. Although these methods generally provide high accurateness and sensitivity, they are also time consuming and require considerable amounts of toxic and expensive chemical solvents and reagents. As an alternative, rapid non-targeted spectroscopic methods have been proposed for food quality control and authentication [15]. Among them, near infrared (NIR), mid infrared (MIR), Fourier transform infrared (FTIR), ultraviolet-visible (UV-Vis) and fluorescence spectroscopies, combined with multivariate approaches, have successfully been applied for the characterization and authentication of vinegars [16,17], their quality control [17,18], and the detection of adulterations [7,19].

The main aim of this work was to investigate the potential of UV-Vis spectroscopy in combination with chemometric tools for discriminating Andalusian PDO wine vinegars according to their geographical origin. Secondly, we aimed to predict the physicochemical and functional properties of vinegars using the spectral information by applying regression analysis.

2. Materials and Methods

2.1. Vinegar Samples

A total of 71 vinegar samples were kindly provided by local wine cellars from the three Andalusian PDOs: 18 "Condado de Huelva" PDO vinegar samples, 8 "Montilla-Moriles" PDO vinegar samples, and 45 "Jerez" PDO vinegar samples. All the samples were kept in darkness at room temperature until analysis.

2.2. Spectral Measurements

Absorption spectra in the ultraviolet-visible (UV-Vis) region were recorded in the range 200–700 nm at wavelength intervals of 2 nm with a scanning speed of 2400 nm min^{-1}, using a Thermo Electron Corporation Spectronic Helios Alpha spectrophotometer (Thermo Scientific™, Waltham, MA, USA). The absorbance was measured using rectangular quartz cuvettes with a path length of 2 mm against deionized water blanks.

2.3. Determination of Chemical Parameters

2.3.1. Total Phenolic Content

The total phenolic content (TPC) of vinegars was measured according to the Folin–Ciocalteu spectrophotometric method [20]. For this purpose, a volume of 0.02 mL of each sample was mixed with 1.58 mL of Milli-Q water and 0.10 mL of the Folin–Ciocalteu reagent (Sigma-Aldrich, Steinheim, Germany). Then, 0.30 mL of 20% sodium carbonate was added, and the mixture was incubated at 40 °C during 120 min using a water bath. Finally, the absorbance was measured at 725 nm using a quartz cuvette against Milli-Q water blanks. The results were expressed as gallic acid equivalents per liter.

2.3.2. Antioxidant Activity

The antioxidant activity of vinegar samples was examined using the method developed by Brand-Williams et al. based on the redox reaction between the radical 1,1-diphenyl-2-picrylhydrazyl (DPPH) and the antioxidants contained in the sample [21]. For this, a total volume of 0.1 mL of the vinegar sample was added to 3.9 mL of 0.1 mM DPPH solution in methanol. The mixture was kept at room temperature for 30 min, and the absorbance was then measured at 515 nm. Complementarily, the ABTS (2,2'-azino-bis(3-ethylbenzothiazoline-6-sulfonic acid)) radical cation decolorization assay was also applied to determine the free-radical scavenging activity of vinegar samples according to the method of Pellegrini et al. [22]. To this end, the ABTS radical cation (ABTS$^{\bullet+}$) was first produced by reacting 7 mM ABTS stock solution with 2.45 mM potassium persulfate at room temperature in darkness for 16 h. Then, the ABTS$^{\bullet+}$ solution was diluted with ethanol to obtain an absorbance of 0.70 ($\pm$0.01) at 734 nm. Finally, a 0.1 mL aliquot of the sample was added to 2.9 mL of the diluted ABTS$^{\bullet+}$ solution, the mixture was incubated for 6 min, and the absorbance was read at 734 nm. The results of both the DPPH and ABTS assays were expressed as Trolox equivalents per liter.

2.3.3. Total Acidity and pH

Total acidity was measured using a Titralyser automatic titrator (Laboratoires Dujardin-Salleron, Noizay, France) according to the Spanish Official Methods for the analysis of vinegars [23]. The results were expressed as grams of acetic acid per 100 mL of vinegar. The pH measurements were performed using the same automatic titrator.

2.4. Data Analysis

Data processing and statistical analyses were performed in the Statistica 8.0 software (StatSoft, Tulsa, OK, USA). To remove undesirable systematic variation in the data due to physical effects, different preprocessing methods were studied prior to multivariate analysis, including standard normal variate (SNV), multiplicative scatter correction (MSC), first derivative (1D) and second derivative (2D). Furthermore, both the raw spectra and the preprocessed spectra were subjected to logarithmic transformation and normalized by autoscaling. Briefly, techniques such as SNV, MSC and autoscaling normalization attempt to reduce spectral scattering, whereas spectral derivatives may help to remove baseline drift, distinguish overlap peaks and extract important signals [24]. Analysis of variance (ANOVA) followed by the Fisher LSD post hoc test was applied to look for differences between the three PDOs under study in terms of the chemical parameters evaluated here (i.e., TPC, antioxidant activity, total acidity, pH). p-Values below 0.05 were considered as statistically significant. Afterwards, linear discriminant analysis (LDA) was employed to build classification models with the aim of assessing the potential of spectroscopy data to authenticate wine vinegars according to the geographical origin. LDA is based on the generation of a number of orthogonal linear discriminant functions equal to the number of categories minus one [25]. Prior to LDA, data normality was checked by inspecting probability plots, and the homogeneity of variance-covariance matrices was tested by applying the Box's M test. Furthermore, the multicollinearity was assessed by using the condition number method, which is defined as the square root of the ratio between the maximum and minimum eigenvalues. The most significant variables involved in sample differentiation according to the PDO were selected using Wilks' λ and F value as criterion for inclusion or removal of variables in the model. Subsequently, the LDA models were subjected to 7-fold cross-validation to assess their predictive ability. For this purpose, the data matrix was randomly divided into two sets, both of them containing the same percentage of samples within each class: a training set that was used to construct the classification model, and a test set to evaluate the model performance. The performance of the models was evaluated by computing their sensitivity (SENS) and specificity (SPEC), where SENS refers to the percentage of cases belonging to a determinate class that were correctly classified, and SPEC refers to the percentage of cases not belonging to a class that were correctly not classified in this class. Finally, partial least squares regression (PLSR) analysis was applied to predict the chemical and functional properties of vinegars from the spectral data. This technique is a quick, efficient and optimal regression method based on covariance, which is highly recommended to avoid overfitting when the number of explanatory variables is large and inter-correlated. This technique is based on building a set of components that accounts for as much as possible variation in the data, while also modeling the Y variables. For this purpose, PLSR works by extracting a set of components that transforms the original X and Y data into a set of t-scores and u-scores, respectively. Then, the t-scores are used to predict the u-scores, which are in turn used to predict the response variables. The statistical performance of these models can be defined by the following parameters: R^2_y, the proportion of variance of the response variable explained by the model; and R^2_x, the proportion of variance in the data explained by the model. Furthermore, their predictive ability was assessed by computing the regression correlation coefficient (R^2), the predicted residual error sum of squares (PRESS) and the residual predictive deviation (RPD). The coefficient of correlation estimates the percentage of variation explained by the model, whereas the PRESS parameter provides a measure of the fit of the regression to a sample of observations that were not used to create the model. The RPD values were computed as the ratio between the standard deviation of the reference values and the error of prediction, so that the higher the RPD values, the greater the probability of the model to accurately predict the chemical parameters. Accordingly, good prediction models require R^2 values to be close to 1, as small as possible PRESS values, and RPD values above 2.5–3.0 [26].

3. Results and Discussion

3.1. Spectral and Chemical Characteristics of Vinegars

The spectral data recorded in this study were in accordance with the characteristic UV-Vis spectra of vinegars reported in the literature [7]. In these spectra, three regions can easily be differentiated in all the vinegar samples regardless of the geographical origin (Figure 1): a strong absorption peak around 200 nm (region of sobresaturation), the absorption band of phenolic compounds in the range 250–400 nm, and the region above 400 nm, where there is practically no absorption.

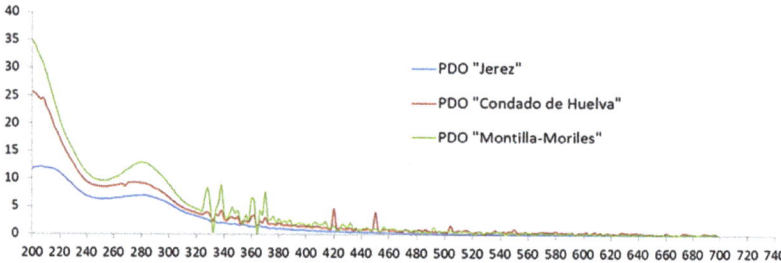

Figure 1. Typical UV-Vis spectra of vinegars.

Furthermore, four parameters related to the physicochemical and functional properties of vinegars were also assessed in this study. The Folin–Ciocalteu method was employed for determining the TPC of vinegars, whereas the antioxidant activity was determined by applying two complementary methods, namely the DPPH and ABTS assays [27]. In addition, the total acidity and pH of the vinegar samples were measured using a titrator. As shown in Table 1, the TPCs and antioxidant activities were within the ranges reported by Kadiroğlu for various types of commercial vinegars [19], whereas total acidity values were similar to those found by De la Haba et al. in vinegars from the PDO "Montilla-Moriles" [16]. In contrast, the pH values were slightly lower than those reported by these same authors. One-way ANOVA revealed significant differences in the four parameters determined here between the samples from the three Andalusian PDOs under study (Table 1). Vinegars from the "Montilla-Moriles" PDO showed a characteristic chemical profile with higher TPC and antioxidant activity, whereas "Condado de Huelva" and "Jerez" samples were characterized by higher total acidity.

Table 1. Mean, minimum and maximum values for the antioxidant activity, total phenolic content, total acidity and pH in vinegar samples from the three Andalusian PDOs, and p-values obtained by ANOVA.

	PDO "Jerez"	PDO "Condado de Huelva"	PDO "Montilla-Moriles"	p-Value
Antioxidant activity—DPPH assay (mmol Trolox equivalents L^{-1})	1.41 [a] (0.30–5.00)	1.39 [a] (0.08–5.94)	5.37 [b] (0.69–16.68)	0.0000
Antioxidant activity—ABTS assay (mmol Trolox equivalents L^{-1})	2.11 [a] (0.29–4.35)	1.45 [a] (0.15–4.53)	15.24 [b] (0.87–52.33)	0.0001
Total phenolic content (mg gallic acid equivalents L^{-1})	450.69 [a] (157.54–1347.56)	984.26 [a] (102.82–9341.47)	2186.14 [b] (206.99–7906.01)	0.0000
Total acidity (g acetic acid 100 mL^{-1})	8.44 [a] (5.58–10.89)	8.72 [a] (6.54–11.04)	7.32 [b] (5.88–10.32)	0.0079
pH	2.03 [a] (1.85–2.42)	1.95 [b] (1.60–2.36)	2.14 [c] (1.60–2.57)	0.0029

Superscript letters within each row indicate significant differences between groups marked with different letters, according to the post-hoc Fisher LSD test ($p < 0.05$).

3.2. Differentiation of Vinegars According to Their Protected Designation of Origin

To evaluate the potential of UV-Vis spectroscopy to differentiate vinegar samples according to the geographical origin, the spectral data recorded here were subjected to linear discriminant analysis (LDA). Furthermore, we also compared the potential of different preprocessing methods to correct data variations that can be caused by physical phenomena (e.g., noise, baseline drift), with the aim of improving the classification performance of the multivariate models.

For stepwise LDA modeling, the original variables were divided into five wavelength intervals, as LDA requires a maximum number of variables equal to the number of cases. Table 2 shows the percentage of correct classifications obtained for each model, the number of components selected, as well as the sensitivity (SENS) and specificity (SPEC) parameters computed by means of cross-validation. The best results were obtained when LDA was carried out on non-preprocessed data in the range 280–400 nm, with 98.6% mean prediction ability (only one "Condado de Huelva" sample was misclassified). This model retained eight components (F to enter = 4.00 and F to remove = 1.00), and the scatter plot of the samples in the plane defined by the two first canonical variables enabled a clear distinction of the vinegar samples according to the PDO. As shown in Figure 2A, the first canonical function differentiated the "Montilla-Moriles" vinegar samples from the other two PDOs, which were in turn separated by the second component. Moreover, the comparison of the different preprocessing techniques demonstrated that SNV-based treatment of the spectral data in the region 502–600 nm also provided good classification performance (88.7%), although the differentiation of the three PDOs in the corresponding scatter plot was not so clear compared to that obtained with non-preprocessed data (Figure 2B). This is in good agreement with previous studies describing the suitability of SNV for the extraction of spectral information related to antioxidants and antioxidant activity by NIR [28], and for the discrimination of Australian Shiraz wines by MIR [29].

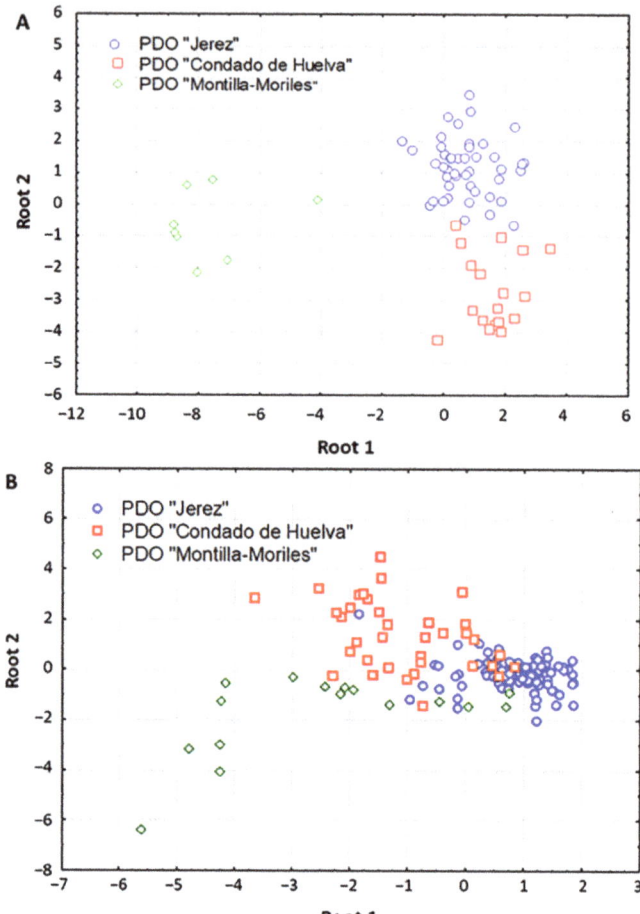

Figure 2. Linear discriminant analysis (LDA) scatter plots showing the distribution of samples in the space defined by the two first canonical variables using: (**A**) non-preprocessed UV-Vis spectral data, (**B**) SNV-preprocessed UV-Vis spectral data.

In a previous study, vinegar samples from "Jerez" and "Condado de Huelva" PDOs were subjected to complete chemical characterization of polyphenols and volatile compounds [13]. Using these chemical descriptors, LDA analysis enabled the differentiation of the samples according to the geographical origin, although yielding lower classification performance to that provided by UV-Vis spectral data (92.86% and 94.12% of the samples were successfully classified when modeling polyphenolic and volatile data, respectively). This therefore highlights the great potential of spectroscopic techniques as simple, ecofriendly and low-cost alternatives against traditional analytical approaches based on chemical analysis for the differentiation of vinegar samples.

Table 2. Statistical performance of the linear discriminant analysis (LDA) models built for the differentiation of vinegars according to the PDO using the UV-Vis spectral data.

Wavelength Interval	Preprocessing Method	Number of Components	Classification Performance	SENS	SPEC
200–278 nm	Raw	4	81.0%	66.0%	83.5%
	SNV	2	76.0%	56.6%	81.6%
	MSC	2	77.0%	60.3%	82.9%
	1D	4	77.0%	58.2%	83.7%
	2D	7	77.5%	45.6%	74.5%
280–400 nm	Raw	8	98.6%	77.8%	89.4%
	SNV	8	83.8%	59.6%	84.7%
	MSC	9	83.0%	61.1%	85.7%
	1D	7	78.0%	44.8%	74.1%
	2D	3	68.3%	43.5%	70.8%
402–500 nm	Raw	3	69.0%	45.5%	73.9%
	SNV	7	82.3%	62.0%	84.0%
	MSC	7	80.0%	59.1%	84.0%
	1D	3	68.0%	54.4%	77.7%
	2D	6	71.8%	37.5%	70.6%
502–600 nm	Raw	3	71.0%	48.3%	75.4%
	SNV	10	88.7%	63.7%	85.3%
	MSC	11	85.9%	61.7%	84.6%
	1D	5	71.8%	42.5%	69.9%
	2D	4	67.6%	42.2%	70.9%
602–698 nm	Raw	11	81.7%	45.0%	74.1%
	SNV	2	80.0%	59.2%	82.4%
	MSC	5	80.0%	60.7%	83.8%
	1D	2	67.6%	46.3%	72.5%
	2D	2	66.9%	41.3%	70.0%

3.3. Prediction of Chemical Parameters of Vinegars Using Spectroscopic Data

To investigate the potential of spectroscopy techniques for predicting the chemical and functional characteristics of vinegar samples, multivariate partial least squares regression (PLSR) was applied to model the linear relationships between the whole UV-Vis spectral data and the chemical parameters under study (i.e., TPC, antioxidant activity, total acidity and pH). As shown in Table 3 and Figure 3, the spectroscopic data demonstrated excellent capacity to predict the antioxidant activity and TPC of vinegars, as expected considering the strong absorption band of phenolic compounds in the UV-Vis region. The best model for predicting the antioxidant activity measured through the DPPH assay was obtained when using non-preprocessed spectral data, which was constructed with three PLS factors accounting for 91.5% and 84.5% of the variability for X and Y variables, respectively ($R^2 = 0.849$, PRESS = 0.561, RPD = 6.95). Similarly, the spectral data also accurately predicted the antioxidant activity determined by means of the ABTS assay, in that case when applying MSC preprocessing ($R^2 = 0.990$, PRESS = 0.220, RPD = 13.18). These results are in great accordance with those reported by Kadiroğlu, who used FTIR spectral data to predict the antioxidant activities of commercial vinegars [19]. With regard to the TPC, the best predictive model was also obtained after applying MSC preprocessing to the spectral data

(R^2 = 0.744, PRESS = 0.715, RPD = 2.75), in line with a previous study describing the application of NIR spectroscopy to characterize "Montilla-Moriles" PDO vinegars [16]. In this respect, it should be noted that regression analysis between non-processed spectral data and TPC provided a higher RPD value to that obtained when using MSC-preprocessed data (Table 3), but the R^2 parameter was below the recommendations by Tamaki and Mazza for accurate predictions [26].

Table 3. Statistical performance of the partial least squares regression (PLSR) models built for the prediction of the chemical properties of vinegars using the UV-Vis spectral data.

Wavelength Interval	Preprocessing Method	Number of Components	R^2_Y	R^2_X	R^2	PRESS	RPD
Antioxidant activity (DPPH assay)	Raw	3	0.845	0.915	0.849	0.561	6.95
	SNV	5	0.469	0.918	0.815	0.773	5.04
	MSC	5	0.894	0.906	0.870	0.703	5.54
	1D	1	0.727	0.318	0.818	1.132	3.44
	2D	9	0.970	0.809	0.845	0.582	6.69
Antioxidant activity (ABTS assay)	Raw	8	0.992	0.973	0.983	0.306	9.50
	SNV	6	0.969	0.909	0.985	0.455	6.37
	MSC	2	0.878	0.838	0.990	0.220	13.18
	1D	1	0.944	0.625	0.945	0.266	10.91
	2D	1	0.953	0.535	0.951	0.236	12.26
Total phenolic content	Raw	2	0.459	0.885	0.456	0.563	3.45
	SNV	2	0.528	0.755	0.659	0.709	2.78
	MSC	2	0.527	0.769	0.744	0.715	2.75
	1D	1	0.516	0.447	0.526	0.687	2.87
	2D	2	0.605	0.559	0.490	0.882	2.23
Total acidity	Raw	1	0.019	0.665	0.053	1.070	1.41
	SNV	6	0.483	0.921	0.219	1.031	1.15
	MSC	6	0.507	0.917	0.394	1.040	1.45
	1D	2	0.109	0.639	0.067	1.038	1.45
	2D	1	0.117	0.189	0.084	1.098	1.37
pH	Raw	4	0.214	0.943	0.074	0.709	0.24
	SNV	4	0.274	0.877	0.183	0.642	0.26
	MSC	6	0.505	0.924	0.384	0.659	0.26
	1D	2	0.111	0.671	0.070	0.683	0.25
	2D	3	0.415	0.605	0.071	0.676	0.25

In contrast, PLSR models for total acidity and pH showed poor predictive ability regardless of the preprocessing technique applied, with R^2 values below 0.4 for both chemical parameters, and RPD values below 1.45 and 0.26 for total acidity and pH, respectively. These results were, however, not surprising since the principal acidity-related compounds present in vinegars do not absorb in the UV-Vis region.

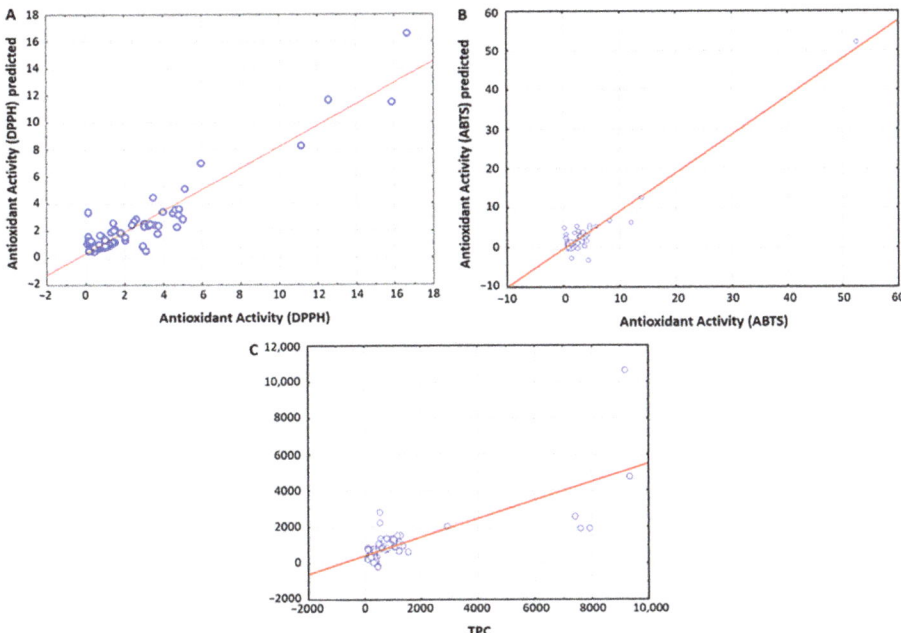

Figure 3. Partial least squares regression (PLSR) analysis for predicting the antioxidant activity measured through the DPPH assay (**A**) and the ABTS assay (**B**), as well as the total phenolic content (**C**) of vinegars using UV-Vis spectral data.

4. Conclusions

In this study, we have demonstrated that UV-Vis spectroscopy in combination with chemometrics tools can be used for the authentication of wine vinegars from Andalusian PDOs, namely "Vinagre del Condado de Huelva", "Vinagre de Montilla-Moriles" and "Vinagre de Jerez". This novel spectroscopic method represents a simple, ecofriendly and low-cost alternative against traditional analytical approaches based on chemical analysis. In particular, LDA modeling of the spectral data recorded within the range 280–400 nm, where phenolic compounds show their characteristic absorption band, enabled a clear differentiation of the three PDOs under study with excellent classification performance (98.6%). Furthermore, PLS regression analysis demonstrated the capacity of UV-Vis spectral data for predicting the TPC and antioxidant activity of vinegars. Altogether, the present study represents one-step further on the development of fast-screening methods for quality control and traceability of wine vinegars. Future studies involving a higher number of vinegar samples are needed to validate the results and conclusions presented here, thus enabling the implementation of this methodology for routine analysis in vinegar certification.

Author Contributions: Conceptualization, Á.F.-R.; methodology, A.S. and Á.F.-R.; validation, Á.F.-R.; formal analysis, R.G.-D., A.S. and Á.F.-R.; investigation, R.G.-D., A.S. and Á.F.-R.; resources, A.S. and Á.F.-R.; data curation, Á.F.-R.; writing—original draft preparation, Á.F.-R.; writing—review and editing, R.G.-D., A.S. and Á.F.-R.; visualization, Á.F.-R.; supervision, Á.F.-R. All authors have read and agreed to the published version of the manuscript.

Funding: This research received no external funding.

Institutional Review Board Statement: Not applicable.

Informed Consent Statement: Not applicable.

Data Availability Statement: The data presented in this study are available on request from the corresponding author.

Acknowledgments: The authors thank to the Protected Designations of Origin "Vinagre de Jerez", "Vinagre del Condado de Huelva" and "Vinagre de Montilla-Moriles" for kindly providing the vinegar samples employed in this study.

Conflicts of Interest: The authors declare no conflict of interest.

References

1. Ho, C.W.; Lazim, A.M.; Fazry, S.; Zaki, U.; Lim, S.J. Varieties, production, composition and health benefits of vinegars: A review. *Food Chem.* **2017**, *221*, 1621–1630. [CrossRef]
2. Ruiz-Muñoz, M.; Cordero-Bueso, G.; Benítez-Trujillo, F.; Martínez, S.; Pérez, F.; Cantoral, J.M. Rethinking about flor yeast diversity and its dynamic in the "criaderas and soleras" biological aging system. *Food Microbiol.* **2020**, *92*, 103553. [CrossRef]
3. The Council of the European Union. *Council Regulation (EC) No 510/2006 of on the Protection of Geographical Indications and Designations of Origin for Agricultural Products and Foodstuffs*. Official Journal of the European Union. 2006, pp. 12–25. Available online: https://eur-lex.europa.eu/LexUriServ/LexUriServ.do?uri=OJ:L:2006:093:0012:0025:en:PDF (accessed on 26 May 2021).
4. The European Comission. REGLAMENTO DE EJECUCIÓN (UE) No 984/2011 DE LA COMISIÓN de 30 de Septiembre de 2011 por el Que se Inscribe Una Denominación en el Registro de Denominaciones de Origen Protegidas y de Indicaciones Geográficas Protegidas [Vinagre del Condado de Huelva (DOP)]. Boletín Oficial del Estado. 2011, pp. 7–8. Available online: https://eur-lex.europa.eu/legal-content/ES/TXT/?uri=CELEX:32011R0984 (accessed on 26 May 2021).
5. The European Comission. REGLAMENTO DE EJECUCIÓN (UE) No 985/2011 DE LA COMISIÓN de 30 de septiembre de 2011 por el que se inscribe una denominación en el Registro de Denominaciones de Origen Protegidas y de Indicaciones Geográficas Protegidas [Vinagre de Jerez (DOP)]. Boletín Oficial del Estado. 2011, pp. 9–10. Available online: https://eur-lex.europa.eu/legal-content/ES/TXT/?uri=CELEX%3A32011R0985 (accessed on 26 May 2021).
6. The European Comission. REGLAMENTO DE EJECUCIÓN (UE) 2015/48 DE LA COMISIÓN de 14 de enero de 2015 por el que se inscribe una denominación en el Registro de Denominaciones de Origen Protegidas y de Indicaciones Geográficas Protegidas [Vinagre de Montilla-Moriles (DOP)]. Boletín Oficial del Estado. 2015, pp. 11–16. Available online: https://eur-lex.europa.eu/legal-content/ES/TXT/?uri=CELEX%3A32015R0048 (accessed on 26 May 2021).
7. Torrecilla, J.S.; Aroca-Santos, R.; Cancilla, J.C.; Matute, G. Linear and non-linear modeling to identify vinegars in blends through spectroscopic data. *LWT Food Sci. Technol.* **2016**, *65*, 565–571. [CrossRef]
8. Bakir, S.; Devecioglu, D.; Kayacan, S.; Toydemir, G.; Karbancioglu-Guler, F.; Capanoglu, E. Investigating the antioxidant and antimicrobial activities of different vinegars. *Eur. Food Res. Technol.* **2017**, *243*, 2083–2094. [CrossRef]
9. Budak, N.H.; Aykin, E.; Seydim, A.C.; Greene, A.K.; Guzel-Seydim, Z.B. Functional properties of vinegar. *J. Food Sci.* **2014**, *79*, 757–764. [CrossRef] [PubMed]
10. Paneque, P.; Morales, M.L.; Burgos, P.; Ponce, L.; Callejón, R.M. Elemental characterisation of Andalusian wine vinegars with protected designation of origin by ICP-OES and chemometric approach. *Food Control* **2017**, *75*, 203–210. [CrossRef]
11. Cejudo-Bastante, M.J.; Durán, E.; Castro, R.; Rodríguez-Dodero, M.C.; Natera, R.; García-Barroso, C. Study of the volatile composition and sensory characteristics of new Sherry vinegar derived products by maceration with fruits. *LWT Food Sci. Technol.* **2013**, *50*, 469–479. [CrossRef]
12. Coelho, E.; Genisheva, Z.; Oliveira, J.M.; Teixeira, J.A.; Domingues, L. Vinegar production from fruit concentrates: Effect on volatile composition and antioxidant activity. *J. Food Sci. Technol.* **2017**, *54*, 4112–4122. [CrossRef] [PubMed]
13. Duran-Guerrero, E.; Schwarz, S.M.; Fernández-Recamales, Á.; García-Barroso, C.; Castro, R. Characterization and Differentiation of Spanish Vinegars from Jerez and Condado de Huelva Protected Designations of Origin. *Foods* **2019**, *8*, 341. [CrossRef]
14. Callejón, R.M.; Tesfaye, W.; Torija, M.J.; Mas, A.; Troncoso, A.M.; Morales, M.L. HPLC determination of amino acids with AQC derivatization in vinegars along submerged and surface acetifications and its relation to the microbiota. *Eur. Food Res. Technol.* **2008**, *227*, 93–102. [CrossRef]
15. Zhang, X.; Yang, J.; Lin, T.; Ying, Y. Food and agro-product quality evaluation based on spectroscopy and deep learning: A review. *Trends Food Sci. Technol.* **2021**, *112*, 431–441. [CrossRef]
16. De la Haba, M.J.; Arias, M.; Ramírez, P.; López, M.I.; Sánchez, M.T. Characterizing and authenticating Montilla-Moriles PDO vinegars using near infrared reflectance spectroscopy (NIRS) technology. *Sensors* **2014**, *14*, 3528–3542. [CrossRef]
17. Ríos-Reina, R.; Elcoroaristizabal, S.; Ocaña-González, J.A.; García-González, D.L.; Amigo, J.M.; Callejón, R.M. Characterization and authentication of Spanish PDO wine vinegars using multidimensional fluorescence and chemometrics. *Food Chem.* **2017**, *230*, 108–116. [CrossRef] [PubMed]
18. Peng, T.Q.; Yin, X.L.; Sun, W.; Ding, B.; Ma, L.A.; Gu, H.W. Developing an Excitation-Emission Matrix Fluorescence Spectroscopy Method Coupled with Multi-way Classification Algorithms for the Identification of the Adulteration of Shanxi Aged Vinegars. *Food Anal. Methods* **2019**, *12*, 2306–2313. [CrossRef]
19. Kadiroğlu, P. FTIR spectroscopy for prediction of quality parameters and antimicrobial activity of commercial vinegars with chemometrics. *J. Sci. Food Agric.* **2018**, *98*, 4121–4127. [CrossRef]
20. Folin, O.; Ciocalteu, V. On tyrosine and tryptophane determinations in proteins. *J. Biol. Chem.* **1927**, *73*, 627–650. [CrossRef]
21. Brand-Williams, W.; Cuvelier, M.E.; Berset, C. Use of a Free Radical Method to Evaluate Antioxidant Activity. *LWT Food Sci. Technol.* **1995**, *28*, 25–30. [CrossRef]

22. Pellegrini, N.; Re, R.; Yang, M.; Rice-Evans, C.A. Screening of dietary carotenoids and carotenoid-rich fruit extracts for antioxidant activities applying the ABTS$^+$ radical cation decolorization assay. *Methods Enzymol.* **1999**, *299*, 379–389.
23. Ministerio de la Presidencia. *Métodos Oficiales de Análisis de Aceites y Grasas*; Cereales y Derivados; Productos Lácteos y Productos Derivados de la Uva (Spanish Official Analytical Methods for Oils and Fats; Cereals; Dairy Products and Enological Products); BOE num. 167; de 14 de Julio de 1977; Boletín Oficial del Estado: Madrid, Spain, 1977; pp. 15800–15808.
24. Cozzolino, D.; Cynkar, W.U.; Shah, N.; Smith, P. Multivariate data analysis applied to spectroscopy: Potential application to juice and fruit quality. *Food Res. Int.* **2011**, *44*, 1888–1896. [CrossRef]
25. Berrueta, L.A.; Alonso-Salces, R.M.; Héberger, K. Supervised pattern recognition in food analysis. *J. Chromatogr. A* **2007**, *1158*, 196–214. [CrossRef]
26. Tamaki, Y.; Mazza, G. Rapid determination of lignin content of straw using Fourier transform mid-infrared spectroscopy. *J. Agric. Food Chem.* **2011**, *59*, 504–512. [CrossRef] [PubMed]
27. Amorati, R.; Valgimigli, L. Advantages and limitations of common testing methods for antioxidants. *Free Radic. Res.* **2015**, *49*, 633–649. [CrossRef] [PubMed]
28. Tahir, H.E.; Xiaobo, Z.; Jiyong, S.; Mariod, A.A.; Wiliam, T. Rapid Determination of Antioxidant Compounds and Antioxidant Activity of Sudanese Karkade (Hibiscus sabdariffa L.) Using Near Infrared Spectroscopy. *Food Anal. Methods* **2016**, *9*, 1228–1236. [CrossRef]
29. Riovanto, R.; Cynkar, W.U.; Berzaghi, P.; Cozzolino, D. Discrimination between Shiraz Wines from Different Australian Regions: The Role of Spectroscopy and Chemometrics. *J. Agric. Food Chem.* **2011**, *59*, 10356–10360. [CrossRef] [PubMed]

Article

A Methodology Based on FT-IR Data Combined with Random Forest Model to Generate *Spectralprints* for the Characterization of High-Quality Vinegars

José Luis P. Calle [1], Marta Ferreiro-González [1,*], Ana Ruiz-Rodríguez [1], Gerardo F. Barbero [1], José Á. Álvarez [2], Miguel Palma [1] and Jesús Ayuso [2]

[1] Department of Analytical Chemistry, Faculty of Sciences, University of Cadiz, Agrifood Campus of International Excellence (ceiA3), IVAGRO, 11510 Puerto Real, Spain; joseluis.perezcalle@uca.es (J.L.P.C.); ana.ruiz@uca.es (A.R.-R.); gerardo.fernandez@uca.es (G.F.B.); miguel.palma@uca.es (M.P.)
[2] Department of Physical Chemistry, Faculty of Sciences, Institute of Biomolecules (INBIO), University of Cadiz, 11510 Puerto Real, Spain; joseangel.alvarez@uca.es (J.Á.Á.); jesus.ayuso@uca.es (J.A.)
* Correspondence: marta.ferreiro@uca.es; Tel.: +34-956-01-6359

Citation: Calle, J.L.P.; Ferreiro-González, M.; Ruiz-Rodríguez, A.; Barbero, G.F.; Álvarez, J.Á.; Palma, M.; Ayuso, J. A Methodology Based on FT-IR Data Combined with Random Forest Model to Generate *Spectralprints* for the Characterization of High-Quality Vinegars. *Foods* 2021, 10, 1411. https://doi.org/10.3390/foods10061411

Academic Editor: Raúl González-Domínguez

Received: 22 May 2021
Accepted: 16 June 2021
Published: 18 June 2021

Publisher's Note: MDPI stays neutral with regard to jurisdictional claims in published maps and institutional affiliations.

Copyright: © 2021 by the authors. Licensee MDPI, Basel, Switzerland. This article is an open access article distributed under the terms and conditions of the Creative Commons Attribution (CC BY) license (https://creativecommons.org/licenses/by/4.0/).

Abstract: Sherry wine vinegar is a Spanish gourmet product under Protected Designation of Origin (PDO). Before a vinegar can be labeled as Sherry vinegar, the product must meet certain requirements as established by its PDO, which, in this case, means that it has been produced following the traditional solera and criadera ageing system. The quality of the vinegar is determined by many factors such as the raw material, the acetification process or the aging system. For this reason, mainly producers, but also consumers, would benefit from the employment of effective analytical tools that allow precisely determining the origin and quality of vinegar. In the present study, a total of 48 Sherry vinegar samples manufactured from three different starting wines (Palomino Fino, Moscatel, and Pedro Ximénez wine) were analyzed by Fourier-transform infrared (FT-IR) spectroscopy. The spectroscopic data were combined with unsupervised exploratory techniques such as hierarchical cluster analysis (HCA) and principal component analysis (PCA), as well as other nonparametric supervised techniques, namely, support vector machine (SVM) and random forest (RF), for the characterization of the samples. The HCA and PCA results present a clear grouping trend of the vinegar samples according to their raw materials. SVM in combination with leave-one-out cross-validation (LOOCV) successfully classified 100% of the samples, according to the type of wine used for their production. The RF method allowed selecting the most important variables to develop the characteristic fingerprint ("spectralprint") of the vinegar samples according to their starting wine. Furthermore, the RF model reached 100% accuracy for both LOOCV and out-of-bag (OOB) sets.

Keywords: characterization; Fourier-transform infrared spectroscopy; cluster analysis; Sherry vinegar; spectralprint; random forest; support vector machine

1. Introduction

The production of high-quality vinegar is increasingly important for manufacturers as consumers' demand for a high-quality product presents a growing trend. The quality of vinegar is heavily determined by numerous factors, among which it is worth noting the raw material, the acetification system, and, in some cases, the specific wooden casks used for its aging [1].

In order to preserve and guarantee the quality of certain vinegars associated with specific geographical areas, the European Union recognizes these vinegars with the category of Protected Designation of Origin (PDO) (Council Regulation (EC) 510/2006). Such is the case of Sherry wine vinegar, from the Jerez-Xérès-Sherry, Manzanilla de Sanlúcar, and Vinagre de Jerez PDO region (in SW Spain). The quality of this PDO vinegar is related to the raw material (i.e., the grape variety), the production process, the type of cask used

(America oak barrels), and the aging method [2]. This gourmet-grade wine vinegar is produced from high-quality Sherry wines which are, in turn, protected by a PDO that establishes very specific and traditional aging methods [3]. Both production and quality are precisely described and strictly regulated by law [4]. Sherry vinegar elaboration consists mainly of two production steps. The first step consists of the acetification procedure, which can be performed by traditional (oak barrels for several months) or industrial methods (steel tanks in just a few hours). Regardless of the acetification procedure applied, vinegar is in every case subjected to aging in oak barrels as the second production step. According to European Regulations, there are three categories of PDO Sherry vinegars depending on their aging time in oak wood barrels as follows: Vinagre de Jerez (at least 6 months of aging time), Vinagre de Jerez Reserva (at least 2 years of aging time), and Vinagre de Jerez Gran Reserva (at least 10 years of aging time) [5]. Vinegar diversity, increasing demand, and the fact that a convincing and objective authentication method is still an unresolved issue, the development of reliable analytical methods that allow establishing valid criteria regarding quality, origin, and verification of the production processes, such as aging, are required.

A large number of regular analytical methods have been developed until now for the characterization of different vinegar types on the basis of determining some of the individual compounds of interest that can be found in vinegar.

Since its aroma profile is considered one of the most important quality indicators of a particular vinegar, gas chromatography/mass spectrometry (GC–MS) continues to be the most widely employed technique for vinegar characterization and quality control [6]. In this sense, Pizarro et al. [7] characterized the volatile content in a number of vinegars to differentiate them according to raw material and production process (with or without aging in wood). Likewise, Marrufo-Curtido et al. employed stir bar sorptive extraction coupled to GC–MS (SPME–GC–MS) to characterize different vinegar samples—including Sherry vinegars—on the basis of the identification of certain volatile compounds, and they succeeded in classifying them by raw material and aging process [5]. Cejudo-Bastante et al. [8] performed a comparative study on the production of vinegar according to the acetification process used. For that purpose, they determined the vinegar samples' polyphenolic and volatile profiles by gas and liquid chromatographic techniques. Ríos-Reina et al. compared three different sampling methods prior to analysis by GC–MS with the aim of determining the more suitable method for the characterization and differentiation between vinegar PDOs and other categories [6].

Other methods such as inductively coupled plasma optical emission spectroscopy (ICP-OES) were also successfully used to characterize the mineral composition of a number of PDO Andalusian wine vinegars and classify them according to their origin [9]. Furthermore, ^{1}H-NMR combined with pattern recognition analysis was successfully used to classify vinegars and wines from different raw materials [10,11].

Although all these methods achieved good results, they also present some drawbacks, since they all require costly equipment, long analysis time, and highly qualified personnel. Although such methods can be considered as perfectly suitable for research purposes, they are deemed otherwise within a regular winery environment. Consequently, spectroscopic methods are becoming more popular for the control of vinegar processes at an industrial scale, since they are rapid, nondestructive, and easily applied in situ. In addition, they require minimal or hardly any sample preparation. These attributes make spectroscopic techniques a more appropriate option for the development of vanguard/rearguard analytical strategies for the control of production processes, so that the need for specific corrective actions can be determined in the shortest possible time [12]. Numerous methods can be found in the literature where the individual identification of the compounds of interest in wine, as well as in related drinks, is conducted by Fourier-transform infrared spectroscopy (FT-IR) [13,14]. It must be noted that the IR spectra from wines, vinegar, and other beverages are complex mixtures of overlapped peaks, i.e., unresolved peaks. Therefore, they cannot be used as regular spectroscopic methods; instead, the application of multivariate regression techniques is mandatory. Individual compounds can be determined; however, no specific signals are used. Instead, a model including many different signals must be used.

However, although most of these spectroscopic methods are based on the identification of individual chemical compounds, in order to ensure vinegar authenticity, several compounds would have to be quantified. On the other hand, more sophisticated fake production methods are being continuously developed, which means that certain minor variations that can make a substantial difference in certain cases represent an increasingly demanding challenge. Thus, the characterization of samples based on a limited number of markers can be sometimes complicated and time-consuming. For this reason, nontargeted chemical analyses based on spectroscopic techniques combined with chemometrics are becoming more frequently used for food characterization [15,16]. In this sense, the use of the whole spectral range in combination with chemometrics to identify a unique fingerprint, which has been recently given the name of spectralprint, allows a rapid characterization of each sample with minimum or no preparation at all [13]. In this sense, Ríos-Reina et al. applied fluorescence excitation–emission spectroscopy coupled to parallel factor analysis (PARAFAC) and support vector machine (SVM) to characterize and classify the three abovementioned Spanish wine vinegars with PDO and concluded that SVM classification models provide higher predictive accuracy (over 92%) [17]. The same authors demonstrated the effectiveness of other spectroscopic techniques, namely, near-infrared spectroscopy (NIRS) in combination with certain chemometrics such as principal component analysis (PCA) and partial least squares - discriminant analysis regression (PLS-DA) regarding a rapid and reliable classification and authentication of Spanish PDO wine vinegars [18]. However, to date, only a few papers have been published regarding the use of FT-IR in combination with pattern recognition techniques to discriminate wine vinegars according to their origins. In this sense, the capacity of FT-IR in combination with PLS for the characterization of Sherry wine according to its aging process was previously studied [19]. Guerrero et al. proved that mid-IR spectroscopy combined with multivariate chemometric techniques could be successfully used to classify vinegar samples elaborated from different raw materials, including apple, white/red wines, or balsamic vinegars, as well as their production processes (with and without aging in wood), with 89% accuracy [20]. Ríos-Reina et al. also studied the potential of FT-IR for the characterization of vinegar and classified them according to the standardized aging categories of high-quality wine vinegars [2]. A reduced number of studies can be found in the literature that investigated the use of the whole ultraviolet/visible (UV/Vis) spectra for vinegar authentication [21].

According to the recently published review by Ríos-Reina et al., there are not many studies where the suitability of the different spectroscopic techniques that have been used in this field were compared for effectiveness [13]. Thus, the development of an analytical method that is suitable to be implemented as a routine for the characterization of vinegar remains a challenge. The same review reported that parametric techniques such as PCA, partial least squares (PLS), PLS-DA, and linear discriminant analysis (LDA) are the most often used chemometric tools. However, the implementation of nonparametric techniques such as SVM or, more specifically, random forest (RF) to the characterization of wine or vinegar is scarce or even nonexistent. Nevertheless, according to several studies with a similar approach, the use of nonparametric techniques has been reported to provide better results [22,23].

Therefore, the aim of this work was to design an analytical method based on FT-IR spectral profiles in combination with support vector machine and random forest to produce *spectralprints* that allow the classification of Sherry wine vinegar according to their starting wine.

2. Materials and Methods

2.1. Samples

A total of 48 samples obtained from different Sherry vinegars under Protected Designation of Origin (PDO) from a local winery (Bodegas Páez Morilla S.A., Jerez de la Frontera,

Spain). All of the samples were "Reserva" vinegars, i.e., more than 2 years of aging in oak barrels. The samples were taken directly from certain oak barrels in the winery. The vinegar samples were divided into three categories accordingly to the origin of the wine used to elaborate the vinegar as follows: 24 Palomino Fino vinegar, 12 Moscatel vinegar, and 12 Pedro Ximénez vinegar. In order to ensure a wide assortment, the samples from the winery were taken directly from oak barrels of varied volume located inside different buildings, at different positions, and containing different starting grape/wines. The samples were tagged after their specific starting wine as follows: PF for Palomino Fino, MO for Moscatel, and PX for Pedro Ximénez, followed by the barrel row level indicated by OB1 for solera (ground level) and OB2 for first criadera (first level)—corresponding to different aging times. In order to reduce turbidity and remove impurities, before being subjected to FT-IR spectrophotometry, the samples were filtered through 0.45 μm filters. No further preparation procedures were required. In addition, all samples were analyzed in duplicate, and the average value for each sample has been used.

2.2. Fourier-Transform Infrared Spectra Acquisition

Fourier-transform infrared spectra were obtained for all samples by means of a MultiSpec (TDI, Barcelona, Spain) spectrophotometer. A 7 mL (standard setting) sample was collected and pumped through the system. The spectra were recorded in the range 952–3070 cm^{-1} with 3.86 cm^{-1} resolution and 20 μm optical path length. The operating temperature was maintained at 25 °C. The total process time per sample was 1 min. All of the samples were measured after doing a blank using a commercial solution provided by TDI, which is a water solution of Triton® (TDI, Barcelona, Spain).

2.3. Data Analysis

No data pretreatment was performed. Thus, the spectral raw data were placed into $D_{n \times p}$ matrices where n denotes the number of samples and p denotes the number of variables. Therefore, a $D_{48 \times 555}$ (48 spectra recorded at 555 different wavenumbers) matrix was obtained for multivariate analysis. The processing of the data, using both unsupervised techniques such as hierarchical cluster analysis (HCA) and supervised nonparametric techniques such as support vector machine (SVM) or random forest (RF), was performed using RStudio software (R version 4.0.5, Boston, MA, USA).

3. Results

3.1. Exploratory Analysis

Firstly, each sample's (n = 48) raw FT-IR spectrum without any pretreatment (p = 555) was subjected to hierarchical cluster analysis (HCA) followed by Ward's method with Manhattan distance to determine any clustering trends. The resulting dendrogram, represented as a phylogenetic tree for easier comprehension, can be seen in Figure 1.

As can be seen, three clearly differentiated branches were obtained—each one corresponding to each of the three starting wines used to elaborate the vinegars. It can be observed that the PX samples (pink color) were closer to the branch containing the MO samples (green color). This suggests that these two types of wine vinegars have a closer similarity with regard to their FT-IR spectrum. Additionally, Figure S1 (Supplementary Materials) shows the FT-IR spectra for all of the vinegar samples, where this greater similarity can be observed. On the other hand, the PF samples (blue color) were the most clearly differentiated from the rest of the samples, especially from the PX ones, which were farther apart. These trends could be associated with the specific ethanol initial content in each starting wine, since PF wines generally exhibit lower ethanol content than MO wines and substantially lower content than PX wines. It must be mentioned that the total acidity of all the vinegars analyzed in this study did not significantly fluctuate.

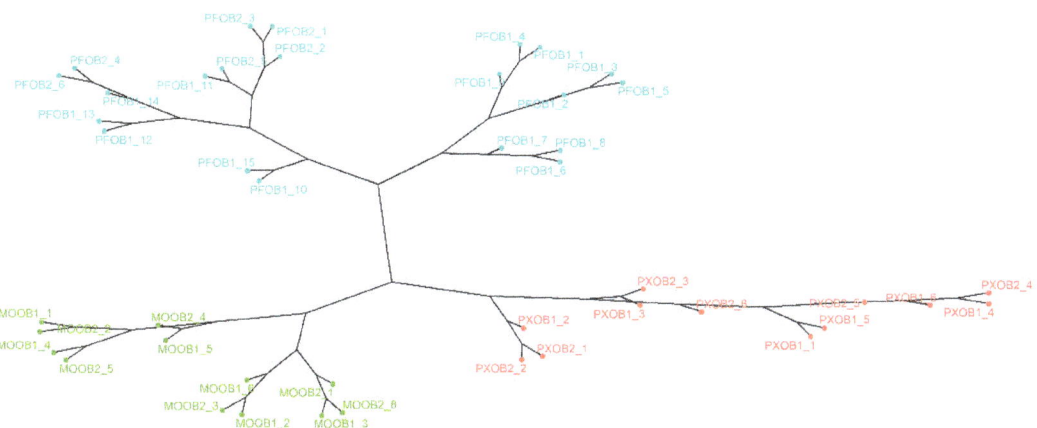

Figure 1. Dendrogram from the HCA analysis combined with Ward's method with Manhattan distance. The vinegar samples are colored according to their starting wine: blue for PF (Palomino Fino), green for MO (Moscatel), and pink for PX (Pedro Ximénez). OB1: solera and OB2: first criadera (n = 48).

It should also be noted that the PF group was the most heterogeneous and that it could be divided into two different sub-branches: one exclusively including samples aged for longer times, i.e., PFOB1, and the other comprising all of the PFOB2 samples plus the remaining PFOB1 samples. These results may suggest that FT-IR could also be somewhat correlated with aging time, even if a clear trend in this sense was not ascertained.

It could be said that, in general, this unsupervised exploratory analysis brought to light some data patterns that would allow differentiating between wine vinegars regardless of their aging time.

Additionally, to corroborate this clustering pattern, as well as the wavenumbers responsible for this trend, principal component analysis (PCA) was carried out. Figure 2A shows the scores obtained by the observations for PC1 and PC2. Figure 2B represents the loadings obtained in each of the PCs. As can be seen, in Figure 2A, the grouping trend was exactly the same as in HCA. In this case, PC1 (explaining 92% of the variability of the data) was mainly responsible for the separation of the samples according to the starting wine vinegar. Thus, PF samples were farther away from the rest, acquiring negative scores for PC1, while MO and PF samples were closer to each other with positive scores. However, the three groups were clearly differentiated. Figure 2B gives an idea of the most important wavenumbers for such a separation. The highest loadings were obtained for the region from 972 to 1174 cm^{-1}, which is related to hydroxyl group (C–O stretching of alcohol). In addition, other spectral regions such as 1600 cm^{-1} (related to aromatic compounds) and 2850 cm^{-1} (related to the O–H stretching of acid components) seem to be important according to the PCA results.

3.2. Supervised Techniques

Both HCA and PCA analyses achieved a quite thorough separation of the vinegar samples according to the type of starting wine. However, this technique does not allow predicting future observations; thus, it is necessary to elaborate a predictive model. For our study, support vector machine and random forest were selected as nonparametric techniques that would allow a predictive model to be generated. According to the most recent studies, RF has never been applied to wine or vinegar samples, while SVM has rarely been used for this purpose [13]. However, both models have exhibited considerable potential when applied to other foodstuffs [22,23].

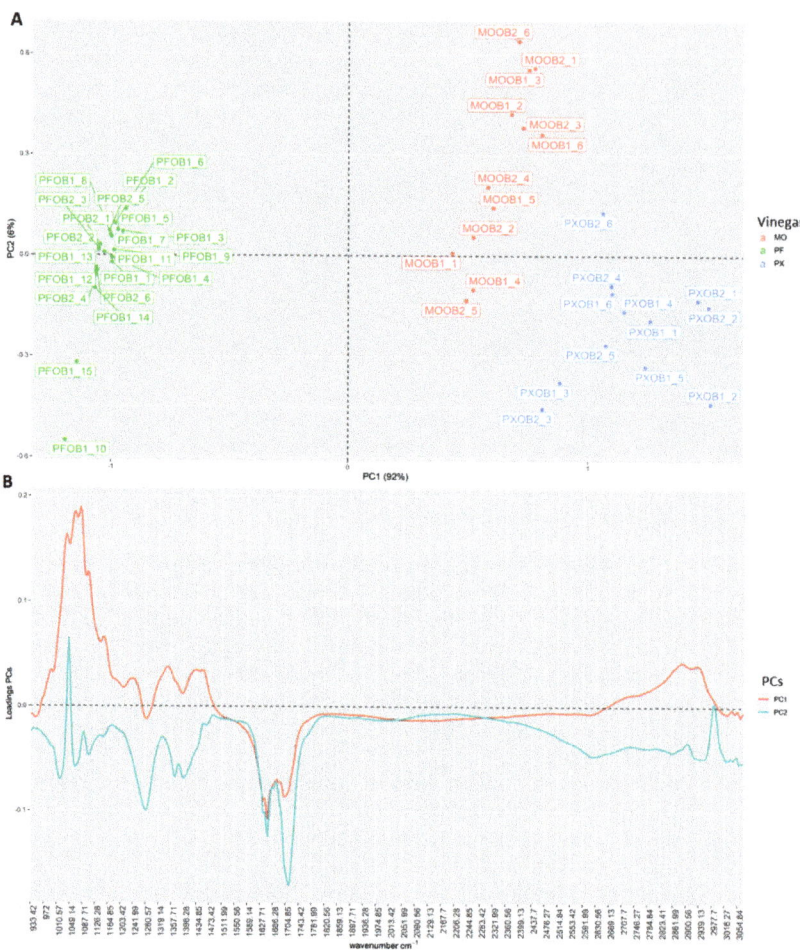

Figure 2. (**A**) Scores obtained by all the samples for the first two principal components (PC1 and PC2); (**B**) loadings obtained in each of the PCs.

3.2.1. Support Vector Machine (SVM)

Support Vector Machine is a supervised method that is commonly used for classification purposes. It is based on a concept known as hyperplane. Hyperplanes allow a clear separation of the items observed according to support vectors. Thus, there is a hyperparameter known as cost (C) that controls the number of support vectors and, consequently, the balance between bias and variance. Furthermore, for this type of analysis, a Radial basis function (RBF) is used so that the separation limits are not linear. Thus, a new hyperparameter known as gamma (γ) is introduced to control the behavior of the Gaussian kernel. Both hyperparameters are to be determined by the analyst and, for this purpose, fivefold cross-validation was used [24,25]. In this case, a grid search method where sequences of C and γ grow exponentially was chosen. Thus, the values in the range (−10, 10) with an increment step of 0.5 units were taken for $\log_2 C$ and $\log_2 \gamma$. Note that, for the fivefold cross-validation, the dataset was divided into five subsets of equal size. Four of such subsets were used to train the model, and the remaining one was used as a test. This process was repeated for each of the subsets. Thus, 8405 models were generated, i.e., 41 × 41 (combinations of C and γ) × 5 (subsets). Figure 3 represents the $\log_2 \gamma$ values (y-axis) versus the log$_2$C values (x-axis) and the accuracy obtained (z-axis). As can be seen, for gamma values roughly below 0.031 (log$_2\gamma = -5$) and cost values above 1 (log$_2$C = 0), the accuracy stabilized at the maximum level. On the one hand, the best results were obtained with the lowest values of gamma and, since this hyperparameter controls the behavior of the kernel, this suggests that the groups were practically linearly separable. In this case, the optimal gamma value was established at 0.00781 (log$_2\gamma = -7$). On the other hand, it seems that the accuracy increased with higher C values. Since this hyperparameter controls the balance between bias and variance, this indicates that fewer misclassified observations would be allowed by the hyperplane and, consequently, there would be fewer support vectors, resulting in a less biased model but with a higher variance. For this reason, the optimal value for C was established at 1 (log$_2$C = 0), since it was the lowest value allowing maximum accuracy. In this way, overfitting (lower variance) was avoided, and excellent performance (lower bias) was achieved, corroborating the robustness of the predictive model. Using the abovementioned hyperparameters, a new model was trained and leave-one-out cross-validation (LOOCV) was performed to determine the error. Both the trained and the LOOCV sets exhibited 100% accuracy, confirming the good performance of the model.

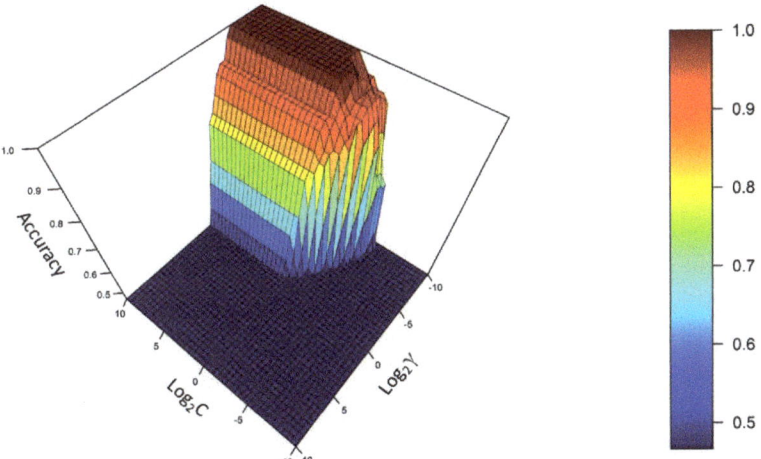

Figure 3. Accuracy of the SVM model calculated using k-fold cross-validation according to log$_2$C and log$_2\gamma$ values.

Although the SVM model proved excellent behavior, the nature of the algorithm does not allow the selection of the most relevant variables regarding the definition of each vinegar *spectralprint* for classification purposes. Consequently, another nonparametric technique known as random forest (RF) was used for that purpose.

3.2.2. Random Forest (RF)

Random forest is a nonparametric supervised technique commonly used for classification and regression purposes. The RF model is made up of multiple individual decision trees trained with a series of random training sets generated by bootstrapping (sampling with replacement). Therefore, there is a data subset, known as out of bag (OOB) which does not contribute to create the model. Thus, in order to evaluate the model performance, a cross-validation method of these OOB instances can be used to determine an unbiased generalization error [26]. Additionally, RF trees are decorrelated by randomly selecting m predictors before evaluating each split in an individual tree. The m value, known as mtry, is a hyperparameter to be optimized by the analyst. For classification purposes, the square root of the total number of predictors is generally set to 24. In addition, a specific number of trees in the RF model must be established. In this sense, a greater number of trees does not result in a greater risk of overfitting, although it should be noted that an excessive number of trees demands longer computation times. Thus, 23 was selected as the mtry value, i.e., the square root of the number of predictors (555 variables). In order to determine the number of trees to be used, models from 2–100 at two-tree intervals were created using the accuracy of the OOB dataset as the criterion (Figure 4). Although the error stabilized rather quickly (at approximately 57 trees), it was continued up to the 100-tree model. In this case, the model with 100 trees was chosen, since it was close to twice the number of trees where the error stabilized.

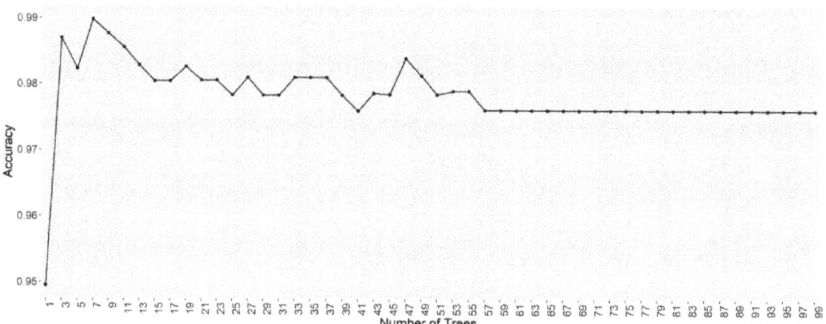

Figure 4. Accuracy of the RF model according to the number of trees.

The model accuracy with the training set was 100%. In addition, two external validations were performed on the OOB and LOOCV sets, with accuracy levels at 97.24% and 100%, respectively. It was, therefore, confirmed that a highly reliable and accurate model was obtained.

Given the nature of the RF model, it is possible to select the most relevant variables for classification purposes. In this case, increasing node purity was the selection criterion.

The six most relevant wavenumbers selected in the RF included some signals related to C–O stretching (1099.28, 1145.57, and 1218.85 cm^{-1}), C–H bending (1457.99 and 1469.56 cm^{-1}), and O–H stretching in a carboxylic acid (2842.7 cm^{-1}). An ANOVA was performed for each of the selected variables, and all of them were significant at a confidence level of 5%.

In order to verify that the model remained stable when based on just those data, a new RF model was elaborated using just those 6 variables. In this case, the value of mtry was set to 2 and the number of trees was set to 100. The result showed 100% accuracy for the training set, as well as for the OOB and LOOCV sets. A summary of the accuracy level achieved by this model (reduced RF) and by the model including all the variables (complete RF) is shown in Table 1. It is worth noting the higher accuracy obtained on the OOB dataset by the reduced RF. This could be explained by the random selection of the samples that make up the OOB set, resulting in hardly any differences between both models. The high collinearity between the predictors in the spectroscopic data is also worth noting. Such collinearity could be explained by the related contributions from the major compounds in the sample, i.e., acetic acid and ethanol, as well as other carboxylic acids and alcohols. Therefore, by reducing from 555 to six variables, this collinearity was reduced and, consequently, accuracy was expected to increase. The possible noise reduction was also expected to contribute to an improvement in the model accuracy. Nevertheless, given the minor error and that collinearity is not so relevant regarding RF models, both of them—complete and reduced models—were considered stable.

Table 1. Accuracy of both RF models for the different validations.

	Accuracy (%)		
	Training Set	**OOB Set**	**LOOCV**
Complete RF	100	97.24	100
Reduced RF	100	100	100

Therefore, the selected variables were used to perform a *spectralprint* that can be used as a suitable routine method for the rapid, reliable, and straightforward identification of wine vinegar types. Figure 5 displays in a radial graph the characteristic *spectralprint* corresponding to each type of wine vinegar. In addition, the mean values of the six variables in each group were normalized to the base peak at 100%. Therefore, intensities and ratios could be used to clearly distinguish each type of wine vinegar. As can be seen, the fingerprint created by each type of vinegar was different.

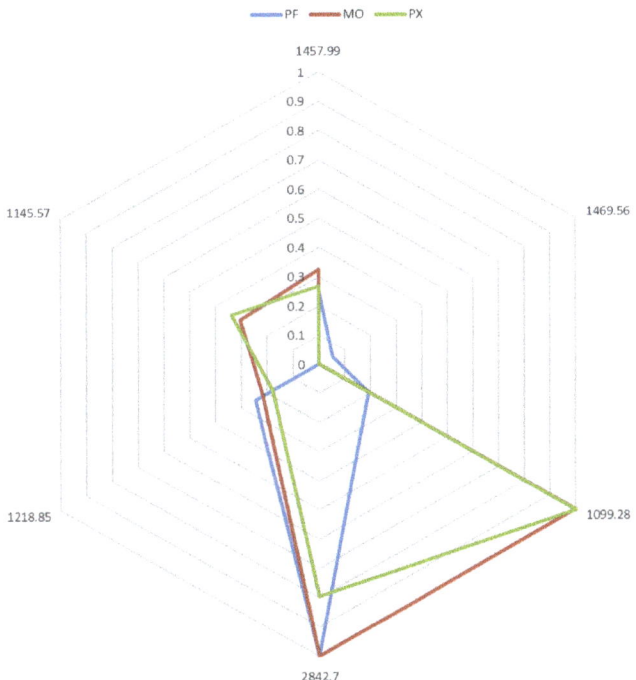

Figure 5. Characteristic *spectralprint* of each wine vinegar type.

PF vinegar, in particular, presented its maximum intensity at wavenumber 2842.7 cm^{-1} and remained below 0.3 at the remaining wavenumbers. This profile was completely different from that of the other groups, especially at wavenumbers 1099.28 cm^{-1} and 1145.57 cm^{-1}. The MO and PX vinegar samples presented more similar profiles, as expected in view of the data from the HCA and PCA analyses, with very high intensities at wavenumber 1099.28 cm^{-1}. However, even if their values at wavenumbers 2842.7, 1218.85, 1145.57, and 1457.99 cm^{-1} diverged, the most notable difference appeared at wavenumber 2842.7 cm^{-1}, where MO vinegar showed its maximum intensity, while PX vinegar reached an intensity around 0.7. The remaining variables, as well as their ratios, were also different for each type of vinegar, thus giving different spectralprints that can be used for the discrimination of the vinegar samples based on the starting wine. This notable signal at 2842.7 cm^{-1} can be attributed to the O–H stretching in carboxylic acid, which suggests that it is similarly and directly related to both prefermentative and fermentative acids, i.e., tartaric, malic, and/or citric acids and succinic, lactic, and acetic acids respectively.

The specific wavenumbers in the spectrum are strongly related to certain specific components associated to the organoleptic properties of vinegar. Thus, some absorptions by hydroxyl groups (C–O stretching of alcohol) were specifically recorded between 1010 cm^{-1} and 1150 cm^{-1}. These absorptions came from ethanol and fusel oils, including propyl alcohol, butyl alcohols, isoamyl alcohols, and hexanol—all of them associated with organoleptic properties. The wavenumbers 1099.28 and 1145.57 cm^{-1}, which were previously identified as the most relevant values with regard to the discrimination of PF vinegars from the other wine vinegars, were within this region. The region between 2840 cm^{-1} and 2940 cm^{-1} presented several absorptions resulting from the O–H stretching of the acid components. The most relevant ones were organic acids such as acetic and tartaric, as well as citric, malic, succinic, and lactic acids. As already mentioned, wavenumber 2842.7 cm^{-1} within this range is quite relevant regarding the discrimination of both PX and MO vinegars. The region between 1450 cm^{-1} and 1510 cm^{-1} describes the absorptions related to aromatic rings (C=C–C stretching). The wavenumbers 1457.99 cm^{-1} and 1469.56 cm^{-1} in this region, which were selected to generate the *spectralprints*, are associated with many of the aromatic compounds in these vinegars. Lastly, the wavenumber 1218.85 cm^{-1} is related to C–O stretching in different kinds of compounds (region from 1200 to 1225 cm^{-1}).

4. Conclusions

FT-IR combined with chemometric tools was demonstrated to be a quite practical methodology for the characterization of Sherry vinegars according to their origin. Specifically, the SVM algorithm applied to the vinegar samples achieved 100% accuracy of the LOOCV. The RF model also displayed an excellent performance at 100% LOOCV and 97.23% OOB accuracy. In addition, the six most relevant wavenumbers were selected from the RF model to create a new RF model that achieved 100% accuracy for both validations (LOOCV and OOB). This is the first time that the RF algorithm has been applied to wine and vinegar samples, and that its validity and robustness have been demonstrated. In addition, the six most relevant wavenumbers were also used to create a characteristic *spectralprint* for each type of wine vinegar, which allowed their rapid, reliable, and uncomplicated differentiation according to their starting wine.

To conclude, spectroscopic techniques were proven to be nondestructive and environmentally friendly methodologies with the capacity to provide rapid and highly reliable results regarding the characterization of vinegars. This, together with their simplicity of use, portability, and low-demanding investment, makes them a highly recommended methodology for in situ routine control of the production and aging processes of vinegars in wineries. In addition, a web platform can be developed with the generated models in order to facilitate data analysis for other users, making the characterization process even easier and more automated.

Supplementary Materials: The following are available online at https://www.mdpi.com/article/10.3390/foods10061411/s1: Figure S1. FT-IR spectrum of all of the samples ($D_{48 \times 555}$). Samples are colored according to the type of wine vinegar: MO (pink), PF (green), and PX (blue).

Author Contributions: Conceptualization, M.F.-G. and M.P.; methodology, J.L.P.C., M.F.-G., and G.F.B.; software, M.F.-G., J.L.P.C., and J.A.; validation, M.F.-G., J.L.P.C., and J.A.; formal analysis, M.F.-G., J.L.P.C., and A.R.-R.; investigation, J.L.P.C. and M.F.-G.; resources, J.A., M.P., and J.Á.Á.; data curation, M.F.-G., J.L.P.C., and A.R.-R.; writing—original draft preparation, M.F.-G. and J.L.P.C.; writing—review and editing, G.F.B. and M.P.; visualization, M.F.-G. and M.P.; supervision, M.F.-G. and M.P.; project administration, J.A., M.P., and J.Á.Á.; funding acquisition, J.A., M.P., and J.Á.Á. All authors read and agreed to the published version of the manuscript.

Funding: This research received no external funding.

Institutional Review Board Statement: Not applicable.

Informed Consent Statement: Not applicable.

Data Availability Statement: The data presented in this study are contained within the article.

Acknowledgments: The authors would like to thank the winery Bodegas Páez Morilla S.A. for providing the Sherry vinegar samples and for the interest shown in the results of this study and Programa de Fomento e Impulso de la Actividad de Investigación y Transferencia de la Universidad de Cádiz for the financial support of this manuscript.

Conflicts of Interest: The authors declare no conflict of interest.

References

1. Jiménez-Sánchez, M.; Durán-Guerrero, E.; Rodríguez-Dodero, M.C.; Barroso, C.G.; Castro, R. Use of ultrasound at a pilot scale to accelerate the ageing of sherry vinegar. *Ultrason. Sonochem.* **2020**, *69*, 105244. [CrossRef] [PubMed]
2. Ríos-Reina, R.; Callejón, R.M.; Oliver-Pozo, C.; Amigo, J.M.; González, D.G. ATR-FTIR as a potential tool for controlling high quality vinegar categories. *Food Control.* **2017**, *78*, 230–237. [CrossRef]
3. Quirós, J.M. La elaboración de vinagre de calidad de Jerez. *Quad. Vitic. Enol. Univ. Torino* **1990**, *2*, 115–129.
4. BOJA 15/10. Inscripción de la Denominación de Origen Protegida "Vinagre de Jerez". Boletín Oficial de la Junta de Andalucía, 2008; Volume 184, pp. 29–35. Available online: https://www.juntadeandalucia.es/organismos/agriculturapescaydesarrollorural/areas/industrias-agroalimentaria (accessed on 1 January 2021).
5. Marrufo-Curtido, A.; Bastante, M.J.C.; Durán-Guerrero, E.; Mejias, R.C.; Natera-Marín, R.; Chinnici, F.; Barroso, C.G. Characterization and differentiation of high quality vinegars by stir bar sorptive extraction coupled to gas chromatography-mass spectrometry (SBSE-GC-MS). *LWT Food Sci. Technol.* **2012**, *47*, 332–341. [CrossRef]
6. Ríos-Reina, R.; Morales, M.L.; González, D.G.; Amigo, J.M.; Callejón, R.M. Sampling methods for the study of volatile profile of PDO wine vinegars. A comparison using multivariate data analysis. *Food Res. Int.* **2018**, *105*, 880–896. [CrossRef]
7. Pizarro, C.; Esteban-Díez, I.; Sáenz-González, C.; Sáiz, J.M.G. Vinegar classification based on feature extraction and selection from headspace solid-phase microextraction/gas chromatography volatile analyses: A feasibility study. *Anal. Chim. Acta* **2008**, *608*, 38–47. [CrossRef] [PubMed]
8. Cejudo-Bastante, C.; Durán-Guerrero, E.; García-Barroso, C.; Mejias, R.C. Comparative study of submerged and surface culture acetification process for orange vinegar. *J. Sci. Food Agric.* **2018**, *98*, 1052–1060. [CrossRef]
9. Paneque, P.; Morales, M.L.; Burgos, P.; Ponce, L.; Callejón, R.M. Elemental characterisation of Andalusian wine vinegars with protected designation of origin by ICP-OES and chemometric approach. *Food Control.* **2017**, *75*, 203–210. [CrossRef]
10. Boffo, E.F.; Tavares, L.A.; Ferreira, M.M.C.; Ferreira, A.G. Classification of Brazilian vinegars according to their ^{1}H NMR spectra by pattern recognition analysis. *LWT Food Sci. Technol.* **2009**, *42*, 1455–1460. [CrossRef]
11. Viskić, M.; Bandić, L.M.; Korenika, A.-M.J.; Jeromel, A. NMR in the Service of Wine Differentiation. *Foods* **2021**, *10*, 120. [CrossRef]
12. Valcárcel, M.; Cárdenas, S. Vanguard-rearguard analytical strategies. *TrAC Trends Anal. Chem.* **2005**, *24*, 67–74. [CrossRef]
13. Ríos-Reina, R.; Camiña, J.M.; Callejón, R.M.; Azcarate, S.M. Spectralprint techniques for wine and vinegar characterization, authentication and quality control: Advances and projections. *TrAC Trends Anal. Chem.* **2021**, *134*, 116121. [CrossRef]
14. Durán, E.; Palma, M.; Natera, R.; Castro, R.; Barroso, C.G. New FT-IR method to control the evolution of the volatile constituents of vinegar during the acetic fermentation process. *Food Chem.* **2010**, *121*, 575–579. [CrossRef]
15. Kansiz, M.; Gapes, J.R.; McNaughton, D.; Lendl, B.; Schuster, K.C. Mid-infrared spectroscopy coupled to sequential injection analysis for the on-line monitoring of the acetone–butanol fermentation process. *Anal. Chim. Acta* **2001**, *438*, 175–186. [CrossRef]
16. Ferreiro-González, M.; Barbero, G.F.; Álvarez, J.A.; Ruiz, A.; Palma, M.; Ayuso, J. Authentication of virgin olive oil by a novel curve resolution approach combined with visible spectroscopy. *Food Chem.* **2017**, *220*, 331–336. [CrossRef] [PubMed]
17. Ríos-Reina, R.; Callejón, R.M.; Savorani, F.; Amigo, J.M.; Cocchi, M. Data fusion approaches in spectroscopic characterization and classification of PDO wine vinegars. *Talanta* **2019**, *198*, 560–572. [CrossRef]
18. Ríos-Reina, R.; Elcoroaristizabal, S.; Ocaña-González, J.A.; González, D.G.; Amigo, J.M.; Callejón, R.M. Characterization and authentication of Spanish PDO wine vinegars using multidimensional fluorescence and chemometrics. *Food Chem.* **2017**, *230*, 108–116. [CrossRef] [PubMed]
19. Ferreiro-González, M.; Ruiz-Rodríguez, A.; Barbero, G.F.; Ayuso, J.; Álvarez, J.A.; Palma, M.; Barroso, C.G. FT-IR, Vis spectroscopy, color and multivariate analysis for the control of ageing processes in distinctive Spanish wines. *Food Chem.* **2019**, *277*, 6–11. [CrossRef] [PubMed]
20. Guerrero, E.D.; Mejias, R.C.; Marín, R.N.; Lovillo, M.P.; Barroso, C.G. A new FT-IR method combined with multivariate analysis for the classification of vinegars from different raw materials and production processes. *J. Sci. Food Agric.* **2010**, *90*, 712–718. [CrossRef]
21. Ríos-Reina, R.; Azcarate, S.M.; Camiña, J.; Callejón, R.M. Assessment of UV–visible spectroscopy as a useful tool for determining grape-must caramel in high-quality wine and balsamic vinegars. *Food Chem.* **2020**, *323*, 126792. [CrossRef] [PubMed]
22. Dankowska, A.; Kowalewski, W. Tea types classification with data fusion of UV-Vis, synchronous fluorescence and NIR spectroscopies and chemometric analysis. *Spectrochim. Acta Part A Mol. Biomol. Spectrosc.* **2019**, *211*, 195–202. [CrossRef] [PubMed]

23. Jia, W.; Liang, G.; Tian, H.; Sun, J.; Wan, C. Electronic Nose-Based Technique for Rapid Detection and Recognition of Moldy Apples. *Sensors* **2019**, *19*, 1526. [CrossRef] [PubMed]
24. Patle, A.; Chouhan, D.S. SVM kernel functions for classification. In Proceedings of the 2013 International Conference on Advances in Technology and Engineering (ICATE), Mumbai, India, 23–25 January 2013; pp. 1–9.
25. Géron, A. *Hands-On Machine Learning with Scikit-Learn and TensorFlow*, 2nd ed.; Roumeliotis, R.N.T., Ed.; O'Reilly Media, Inc.: Sebastopol, CA, USA, 2019.
26. Rodriguez-Galiano, V.F.; Ghimire, B.; Rogan, J.; Olmo, M.C.; Rigol-Sanchez, J.P. An assessment of the effectiveness of a random forest classifier for land-cover classification. *ISPRS J. Photogramm. Remote Sens.* **2012**, *67*, 93–104. [CrossRef]

Article

Discriminant Analysis of the Geographical Origin of Asian Red Pepper Powders Using Second-Derivative FT-IR Spectroscopy

Miso Kim [1,†], Junyoung Hong [1,†], Dongwon Lee [1], Sohyun Kim [1], Hyang Sook Chun [2], Yoon-Ho Cho [3], Byung Hee Kim [4] and Sangdoo Ahn [1,*]

1. Department of Chemistry, Chung-Ang University, Seoul 06974, Korea; rlaalth1328@naver.com (M.K.); hjuny94@hanmail.net (J.H.); idleplanet@naver.com (D.L.); dragon725@naver.com (S.K.)
2. Department of Food Science & Technology, Chung-Ang University, Ansung 17546, Korea; hschun@cau.ac.kr
3. Department of Civil and Environmental Engineering, Chung-Ang University, Seoul 06974, Korea; yhcho@cau.ac.kr
4. Department of Food and Nutrition, Sookmyung Women's University, Seoul 04310, Korea; bhkim@sookmyung.ac.kr
* Correspondence: sangdoo@cau.ac.kr; Tel.: +82-2-820-5230
† These authors contributed equally to this work.

Abstract: This study aimed to discriminate between the geographical origins of Asian red pepper powders distributed in Korea using Fourier-transform infrared (FT-IR) spectroscopy coupled with multivariate statistical analyses. Second-derivative spectral data were obtained from a total of 105 red pepper powder samples, 86 of which were used for statistical analysis, and the remaining 19 were used for blind testing. A one-way analysis of variance (ANOVA) test confirmed that eight peak variables exhibited significant origin-dependent differences, and the canonical discriminant functions derived from these variables were used to correctly classify all the red pepper powder samples based on their origins. The applicability of the canonical discriminant functions was examined by performing a blind test wherein the origins of 19 new red pepper powder samples were correctly classified. For simplicity, the four most significant variables were selected as discriminant indicator variables, and the applicable range for each indicator variable was set for each geographical origin. By applying the indicator variable ranges, the origins of the red pepper powders of all the statistical and blind samples were correctly identified. The study findings indicate the feasibility of using FT-IR spectroscopy in combination with multivariate analysis for identifying the geographical origins of red pepper powders.

Keywords: fourier-transform infrared (FT-IR) spectroscopy; second-derivative spectrum; red pepper powder; geographical origin; discriminant analysis

1. Introduction

Red peppers (*Capsicum annuum* L.) are perennial plants of the family Solanaceae and are widely grown worldwide. The capsaicinoids specifically contained in red peppers are pungent alkaloids and are known to promote energy metabolism [1]. In addition, carotenoids and vitamin C, which are abundant in red peppers, have been reported to have anti-cancer effects [2,3]. Red peppers are mainly used for their hot spicy flavor and red color. They are predominantly processed into a dried powder form for easy transport to markets worldwide. The quality and cost of red pepper powders vary considerably depending on their country of origin. For instance, the quality of imported peppers is reduced owing to freezing or other pretreatment processes [4]. Consequently, consumers typically prefer domestic products [5]. In some cases, retailers deceive consumers by omitting the country of origin of the red pepper powders to inflate their margins [6]. Therefore, it is necessary to develop an accurate and rapid method for identifying the origin of red pepper powders.

Several factors contribute to the differences between plants of different geographical origins [7,8]. Each country has a different crop cultivation environment, such as soil and climate, which lead to differences in the metabolite compositions of plants. Therefore, many studies have been conducted to discriminate foods and agricultural products based on their geographical origins by analyzing their metabolite profiles using mass spectrometry (MS) [9], nuclear magnetic resonance (NMR) spectroscopy [10,11], and chromatography [12]. Recently, simple non-destructive Fourier-transform infrared (FT-IR) spectroscopic techniques have been used for metabolite research [13,14]. The FT-IR method has been applied in combination with multivariate statistical analysis for the geographical discrimination of Korean and Chinese soybeans [15], European saffron [16], European olive oils [17], and European and South American honey [18]. FT-IR, which provides a variety of information on chemical bonds and functional groups, involves the use of inexpensive equipment and does not require the special pretreatment of samples regardless of their state. Diverse FT-IR information has also been used to study the adulteration and authenticity of foods [19]. Moreover, recent studies have combined FT-IR with other instrumental analysis methods to improve the efficiency of food analysis [20,21]; in particular, multivariate statistical analyses have been applied to effectively process instrumental analytical data, requiring the observation of minute differences in the compositions or spectral profiles of various metabolites [22].

The geographical classification of Asian red peppers has also been studied [23–26]. Yin et al. proposed a simple method to determine the geographical origins of red peppers from various regions in China through multivariate analysis of sensory characteristics, such as color, taste, and smell [24]. Zhang et al. conducted a study to effectively distinguish the regions of origin of red peppers in China by analyzing the multi-elements obtained using the inductively coupled plasma (ICP) methods with various chemometric tools [25]. Song et al. demonstrated the possibility of discriminating the geographical origins of Korean red peppers through multivariate statistical analysis of the stable isotope ratio ($^{87}Sr/^{86}Sr$) affected by cultivation environments [26]. Recently, it has been demonstrated that ^1H NMR spectroscopy, combined with multivariate statistical analyses, has a significant predictive potential for identifying the geographical origins of Asian red pepper powders [23]. As is well known, NMR spectroscopy is an extremely beneficial analytical method for metabolite studies because NMR measurements are highly reproducible and can provide both qualitative and quantitative information simultaneously. However, NMR spectroscopy incurs a high instrumental cost and requires pretreatment processes to extract the metabolite components from solid samples, such as red pepper powders.

In this study, FT-IR spectroscopy was used to develop an alternative, simple, and convenient experimental method combined with multivariate statistical analysis for determining the geographical origins of red pepper powders from Korea, China, and Vietnam. The second derivative of the FT-IR spectrum was used to enhance the resolution of broadly overlapping peaks and improve peak quantification by removing baseline errors [27–30]. A one-way analysis of variance (ANOVA) test was used to identify whether the peak variables were significantly different depending on their origins [31]. Canonical discriminant analysis, as a multivariate statistical analysis, was used to obtain the discriminant functions for classifying the geographical origins of red pepper powder samples [32]. The obtained canonical discriminant functions represent linear combinations of the meaningful variables identified by the ANOVA test, and exhibit the highest possible multiple correlations with the origins. The applicability of the discriminant functions was verified through a blind test. Additionally, to easily discriminate new red pepper samples based on their geographical origins without complicated statistical processes, several variables having significant influences on the discrimination were selected as the discriminant indicator variables, and their applicable ranges were set for each geographical origin. The feasibility of this method was also verified by using it to correctly identify the origins of the blind samples. When appropriate discriminant indicator variables and ranges are set, the application becomes very simple. Thus, the proposed method may be considered as a more effective and convenient method for discriminating new samples in comparison with other discrimination methods that require statistical processing.

2. Materials and Methods

2.1. Red Pepper Powder Samples

A total of 105 Asian (Korean, Chinese, and Vietnamese) red pepper powders (or dried red peppers) distributed in Korea were collected as samples. Korean red pepper powders were obtained from local producers or reliable suppliers, such as agricultural cooperatives. Chinese and Vietnamese red pepper powders imported to Korea, through the Korea Agro-Fisheries Trade Corporation, were purchased from local markets. Among the total samples, 86 red pepper powders (Korean = 50, Chinese = 23, Vietnamese = 13) were used for statistical analyses to establish the discriminant functions and indicator variables, which could be used to distinguish their geographical origins. The remaining 19 red pepper powders (Korean = 9, Chinese = 5, Vietnamese = 5) were used as blind test samples to verify the applicability of the established discriminant functions and indicator variables. All the dried red pepper powder samples were stored in a refrigerator at 4 °C. However, commonly sold red pepper powders are mixtures of peel and seed fragments with a length of 1–3 mm, making it difficult to reflect all the component information in the FT-IR spectra and ensure reproducibility of measurements. Therefore, the powder was further ground into a fine powder (with particle diameters of ≤ 200 μm) in a food grinder before measurement. The prepared fine powder samples exhibited good reproducibility in repeated measurements. Three Korean samples purchased in the form of dried red pepper were first ground into powders using a crusher, and then further ground into finer powders as in the other samples.

2.2. FT-IR Measurement

Finely ground red pepper powder samples were loaded onto an FT-IR spectrometer (TENSOR-27; Bruker Optics GmbH, Karlsruhe, Germany) equipped with a diamond attenuated total reflectance (ATR) accessory (A225/Q Platinum ATR; Bruker Optics GmbH, Karlsruhe, Germany). All the spectra were acquired in absorbance mode, in the wavenumber range of 4000–400 cm^{-1}, with 32 repeated scans and a resolution of 4 cm^{-1}. The acquisition time was less than 1 min for each measurement. To ensure the representativeness and reproducibility of the obtained FT-IR spectra, measurements were repeated five times for each sample and statistically averaged. The ATR crystal was cleaned with ethyl alcohol before every measurement. Atmospheric correction was also performed for each measurement to eliminate the effects of CO_2 and H_2O in the atmosphere. OMINC software (version 8.2, ThermoFisher Scientific Inc. Waltham, MA, USA) was used to process the obtained spectra of the red pepper powder samples. Second derivatives of the processed FT-IR absorbance spectra were derived using the Savitzky–Golay (SG) numerical algorithm with third-order polynomials at seven smoothing points [33–35]. The SG method is a commonly used filtering and smoothing technique to remove background effects and any possible noise in the spectrum during second-order differentiation. Through the differentiation process, the sensitivity and resolution of the spectrum were improved by correcting the baseline drift and separating the overlapped peaks [33–35]. The normalized value of each peak in the second-derivative spectra was used for statistical analyses to establish the origin discriminant functions and indicator variables.

2.3. Multivariate Statistical Analysis

Statistical analyses for the second-derivative FT-IR spectral data were performed using IBM SPSS Statistics software (version 26, SPSS Inc., Chicago, IL, USA). Tests of homogeneity of variance were conducted to determine if each peak variable was equally distributed according to the origin group. For the variables with equal variance, a one-way ANOVA test was used to determine significant differences in the peak variables depending on their origin group (significance level, $p < 0.05$). Canonical discriminant analyses were performed with the selected variables to determine the discriminant functions capable of effectively classifying the geographical origins of the red pepper powder samples. Additionally, by selecting several indicator variables that contribute significantly to the discriminant functions and setting the ranges of their values, we determined whether the geographical origin could be easily identified without the statistical dataset. The applicability of both the discriminant functions and the indicator variables obtained were tested through a blind test [28,36–39].

3. Results and Discussion

3.1. FT-IR Spectrum of Red Pepper Powder

Figure 1 illustrates a representative FT-IR spectrum of a red pepper powder sample and its second-derivative spectrum. Using the second derivative of the FT-IR spectrum, more sophisticated spectral data were obtained, while broadly overlapping peaks in the original absorption spectrum could be isolated. The second-derivative process also improved the peak quantification by removing the baseline errors. As summarized in Table 1, 19 distinguishable peaks were selected and labeled in the second-derivative FT-IR spectrum. The peaks were assigned by referring to previous studies [15,16,28,40–45].

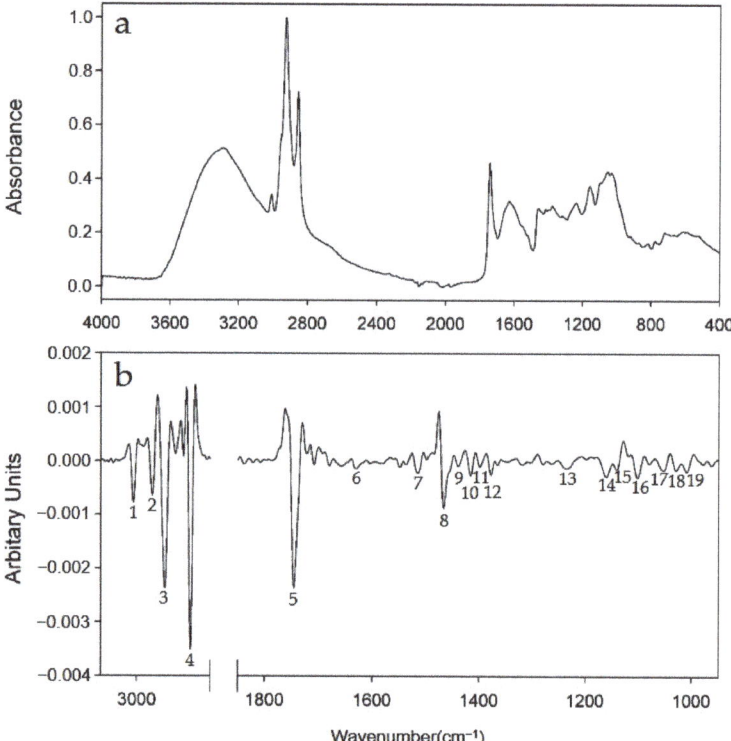

Figure 1. Representative (**a**) FT-IR absorption spectrum of a Korean red pepper powder sample, and (**b**) its second-derivative spectrum (partially expanded at the variables) with numbered peaks (corresponding to the variables in Table 1).

Table 1. Assignment of peaks in the second-derivative FT-IR spectrum of red pepper powders.

Variable	Wavenumber (cm^{-1})	Functional Group	Mode of Vibration
P1	3010	=C–H (cis-)	Stretching [40,41]
P2	2958	–C–H (CH$_3$)	Stretching [40,41]
P3	2924	–C–H (asym CH$_2$)	Stretching [40,41,44]
P4	2852	–C–H (sym CH$_2$)	Stretching [40,41,44]
P5	1745	–C=O (ester)	Stretching [40,41,44]
P6	1653	–C=C– (cis-)	Stretching [40,41]
P7	1516	–C–C– (aromatic)	Stretching [42]
P8	1468	–C–H (CH$_2$, CH$_3$)	Bending [40,41,44]
P9	1439	–C–H (CH$_2$)	Bending [45]
P10	1415	=C–H (cis-)	Bending [28,41,42]
P11	1398	–C–H (CH$_2$, CH$_3$)	Bending [15,16,41]
P12	1377	–C–H (CH$_3$)	Bending [16,40,41,44]
P13	1238	–C–O (ester), –C–H (CH$_2$)	Stretching, Bending [41,44]
P14	1159	–C–O (ester)	Stretching [40,41,44,45]
P15	1142	–C–O	Stretching [43,45]
P16	1101	–C–O	Stretching [43,45]
P17	1053	–C–O	Stretching [43,45]
P18	1028	–C–O	Stretching [43,45]
P19	1008	–C–O	Stretching [43,45]

The broad band at approximately 3400 cm^{-1} is mainly due to the stretching of the O–H bonds, because red pepper powder has a low protein content [7] and easily absorbs moisture [46]. The peak at 3010 cm^{-1} (P1) is attributed to sp^2 C–H stretching, while the peaks at 2958 cm^{-1}, 2924 cm^{-1}, and 2852 cm^{-1} (P2, P3, P4) are attributed to the sp^3 C–H stretching of metabolites in the red pepper powders. The strong peak at 1745 cm^{-1} (P5) is due to the C=O stretching, and the weak peak at 1653 cm^{-1} (P6) is due to the C=C stretching. The aromatic C-C stretching band appears at 1516 cm^{-1} (P7), and several C–H bending bands appear at 1516–1238 cm^{-1} (P8–P13). The various C–O stretching bands of the ester and ether groups appear at 1238–1008 cm^{-1} (P3–P19) which are mainly attributed to the lipids and carbohydrates in the red pepper powders [43,45]. The intensities of the peaks differ slightly depending on the distribution of various metabolites in the red pepper powders. Hence, the statistical analysis of this information could be used to discriminate between the geographical origins. For further statistical analysis, the absolute peak values normalized by the intensity of the C–O stretching peak at 1008 cm^{-1} were used.

3.2. Statistical Analysis

3.2.1. Canonical Discriminant Analysis

Canonical discriminant analysis was performed as a multivariate statistical analysis to achieve the most discriminative peak variables for the arrangement of red pepper powder samples in a lower dimensional space by maximizing the distances between the origin groups. To ensure the robustness of these statistical processes, the homogeneity of the variance of each variable must be considered [31]. Therefore, to select suitable variables for the statistical analysis, a variance homogeneity test was conducted first. As a result of testing 18 peaks, it was confirmed that eight peak variables, namely P5, P7, P8, P10, P12, P14, P16, and P17, had equal variance ($p > 0.05$), while the 10 remaining peaks did not exhibit equal variance ($p < 0.05$) (Table S1).

In this study, an ANOVA test was performed to determine the second-derivative FT-IR peak variables with meaningful differences among the Korean, Chinese, and Vietnamese red pepper powder groups. The ANOVA test verified the equality of the group means of variables using the F test, and determined whether the means of three or more groups were different [31]. Since the ANOVA test is a parametric test, only the eight peaks with equal variance identified in the previous test of homogeneity of variance were considered [31]. All the eight peak variables exhibited significant differences in the origins ($p < 0.001$) with large *F* values (Table 2). As can be seen in Table 2, a smaller Wilks' lambda value (i.e., a larger *F*-value) implies a higher significance in the discrimination analysis.

Table 2. Tests of equality of group means.

Variable	Wilks' Lambda	F	df_1	df_2	Significance Level, *p*
P05	0.480	44.881	2	83	<0.001
P07	0.776	11.946	2	83	<0.001
P08	0.401	61.900	2	83	<0.001
P10	0.550	34.005	2	83	<0.001
P12	0.264	115.938	2	83	<0.001
P14	0.383	66.762	2	83	<0.001
P16	0.623	25.106	2	83	<0.001
P17	0.246	127.075	2	83	<0.001

These eight significant variables were used for the canonical discriminant analysis to establish the discriminant functions. Two canonical discriminant functions were derived for identifying the red pepper powder samples from different origins, and accounted for 100% of the variance. Functions 1 and 2 accounted for 65.2% and 34.8% of the total variance, respectively. The separation between the red pepper powder samples of different geographical origins in the discriminant space was investigated by scatter plotting the discriminant function scores. The score plot showed good separation among the samples from three different origins (Figure 2), suggesting that the variables used to derive the discriminant functions provided sufficient information to identify the geographical origins of red pepper powders. The Korean and Vietnamese samples were found to be completely distinguishable from each other, while the Chinese samples appeared relatively widely scattered between the Korean and Vietnamese samples. This may be attributed to the diversity of the Chinese samples, reflecting the characteristics of China's large geographical area.

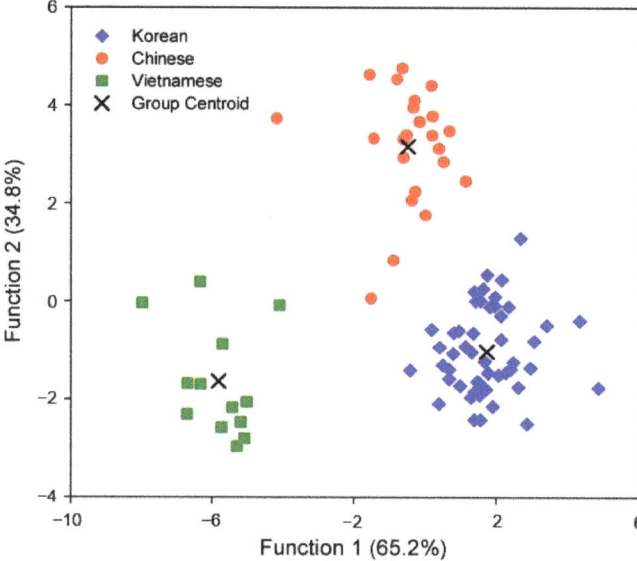

Figure 2. Scatter plot of two discriminant scores for the geographic origins of Korean, Chinese, and Vietnamese red pepper powders.

To verify and examine the predictive discrimination capability of the established canonical discriminant functions, we reclassified the red pepper powder samples used in the multivariate statistical analysis, according to their geographical origins. Table 3 indicates that the canonical discriminant functions correctly classified all 86 red pepper powder samples (50 Korean, 23 Chinese, and 13 Vietnamese) according to their geographical origins (100% of the original group cases were correctly classified), while only one Chinese sample was incorrectly classified in the cross-validation (98.8% of the original group cases were correctly classified). These results were similar to the discrimination results of the origins of 62 Asian red pepper powder (36 Korean, 17 Chinese, and 9 Vietnamese) samples using ^{1}H NMR spectroscopy [23]. In particular, this result was of significance considering that various metabolite components even with minor contents could be used as individual indicators in the ^{1}H NMR analysis. By comparing the analysis results of the mineral elements [25] and sensor characteristics [24] of red peppers from other regions in China using various multivariate statistical analysis methods, it can be observed that their regional scopes were different. However, it can be confirmed that the second-derivative FT-IR method can be sufficiently utilized to discriminate the origins of red pepper powders. In addition, similar discrimination abilities can be confirmed by comparing previous results of the origins of other foods, such as olive oil and honey, using the FT-IR technique [17,18]. Overall, these results indicate that second-derivative FT-IR spectroscopy combined with canonical discriminant analysis has the potential to discriminate Asian red pepper powders according to their geographical origins.

Table 3. Reclassification results for the origins of red pepper powder samples using the canonical discriminant functions.

Origin		Predicted Group			Total	Accuracy (%)
		Korean	Chinese	Vietnamese		
Original	Korean	50	0	0	50	100
	Chinese	0	23	0	23	100
	Vietnamese	0	0	13	13	100
Cross-validated [a]	Korean	50	0	0	50	100
	Chinese	1	22	0	23	95.6
	Vietnamese	0	0	13	13	100

[a] Cross-validation was performed only for those cases in the analysis. In cross-validation, each case is classified using the functions derived from all other cases except that case.

3.2.2. Discriminant Indicator Variables

It was confirmed that Asian red pepper powders could be effectively discriminated according to their geographical origins by canonical discriminant analysis of the signals obtained from the second-derivative FT-IR spectra. This protocol can also be applied to the discrimination of new red pepper powder samples through statistical processes. If several indicator variables suitable for discriminating the origin of red pepper samples are selected and appropriate ranges are set for them, rapid and facile discrimination of the geographical origins of new red pepper powder samples is possible without the need for a specific statistical program or process.

The Pearson coefficients are summarized in the structure matrix table (Table 4), which shows the correlation of each variable with each canonical discriminant function [47–49]. This table reveals that P12 and P17 are the most significant variables in discriminant Functions 1 and 2 (with correlations of −0.475 and 0.714), respectively. P14, and P8 also show high significance in both functions.

Table 4. Structure matrix table with coefficients for the peak variables used in discrimination analysis.

Variable	Structure Matrix	
	Function 1	Function 2
P5	−0.303	0.333
P7	−0.184	0.11
P8	−0.326	0.437
P10	−0.202	0.371
P12	−0.475	0.555
P14	−0.368	0.409
P16	−0.185	0.306
P17	0.394	0.714

These four peak variables (P8, P12, P14, and P17) were also found to have high significance in the mean difference, with an F-value of 60 or more in the one-way ANOVA test (Table 2). The distribution of data between the geographical origin groups of these four variables were compared as box plots (Figure 3), confirming that P12 and P17 were the most effective variables for discriminating the Korean and Vietnamese samples, respectively, from those of other geographical origins. Additionally, the distribution characteristics of P8, P12, and P14 were similar, whereas those for P17 were different. This was also confirmed in the Pearson correlation matrix, which shows the correlations among variables (Table S2).

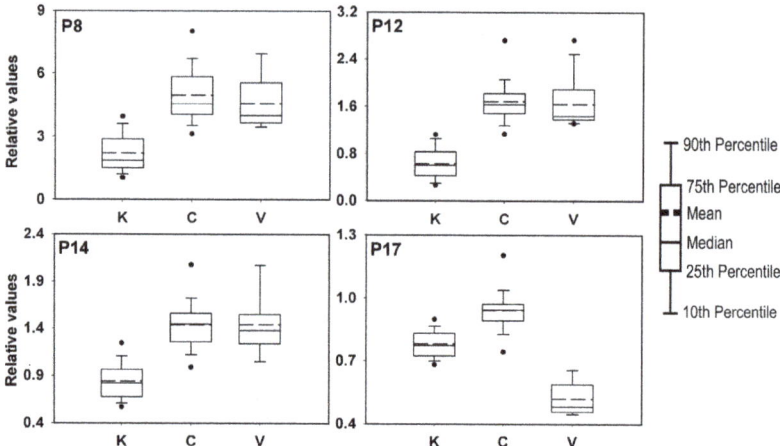

Figure 3. Box plots for variables that are highly correlated with the canonical discriminant functions (K = Korean, C = Chinese, and V = Vietnamese). The dots indicate the 5th and 95th percentiles.

Considering their correlation with the discriminant functions, mean difference, and difference in distribution values, P8 and P14, along with the most significant variables P12 and P17, were selected as indicator variables for discriminating the origins of Asian red pepper powder samples. To discriminate the geographical origins using the specific indicator variables, they must have ranges differentiated according to the origins.

For the Korean red pepper samples, the distribution values of P8 and P12 were smaller than those of the others. These signals can be attributed to C–H stretching vibrations, which are derived from various metabolites containing alkyl groups, and are likely largely influenced by the hydrocarbon chains of fatty acids. Because the fatty acid content is relatively higher in seeds than in the peel of red pepper [50], it can be estimated that the Korean red pepper powder samples contain relatively fewer seeds than the Chinese or Vietnamese samples. Moreover, the P17 signal attributed to the C–O stretching vibration arising mainly from the fructosyl unit [45] was observed to be small in the Vietnamese samples. This implies that the Vietnamese red pepper powders had relatively lower fructose content than those of the Korean and Chinese peppers, which was also confirmed in previous NMR experiments (Figure S1) [23]. For the Chinese red pepper powder samples, all four variables exhibited relatively higher means than the others. However, owing to the diversity of the Chinese samples, the ranges of all the indicator variables significantly overlapped with the ranges of those for other origins; hence, establishing independent variable ranges for Chinese samples was not possible.

Based on these observations, the ranges of the discriminant indicator variables that could discriminate between Korean and Vietnamese red pepper powder samples were set as presented in Table 5.

Table 5. Ranges of the indicator variables for Korean and Vietnamese samples.

Peak No.	Wavelength (cm^{-1})	Vibration	Range	
			Korean	Vietnamese
P8	1468	C–H (CH$_2$, CH$_3$) bending	<4.945	>3.445
P12	1377	–C–H (CH$_3$) stretching	<1.155	>1.305
P14	1159	–C–O (ester) stretching	<1.555	0.985–2.085
P17	1053	–C–O stretching	0.620–0.945	<0.699

The range of each discriminant variable was set based on their maximum or minimum values, or by considering values between the minimum and maximum based on the relative distribution characteristics of each variable value [27,37,38]. For example, in the case of the P8 variable, because Korean red pepper powders had the lowest distribution, its range was set below the maximum value for Korean samples. On the contrary, the Vietnamese samples had a relatively high distribution and, thus, were set above the minimum value for Vietnamese samples. It is worth noting that if each variable value obtained the analysis of more samples satisfied the normal distribution sufficiently, the ranges could be established using a statistical technique as well.

To confirm the suitability of the selected indicator variables and their range settings, we reclassified the red pepper powder samples used in the multivariate statistical analysis, based on their geographical origins. A sample was attributed to a specific origin only if the values of all the indicator variables for the sample were within the discriminant ranges for that origin; the results are summarized in Table 6. When the ranges of the indicator variables for the Korean red pepper powder samples were applied, all 50 Korean samples were identified as "Korean," and the remaining 36 samples (23 Chinese and 13 Vietnamese) were all classified as "not Korean." When applying the ranges of the indicator variables for the Vietnamese red pepper powder samples to the 36 "not Korean" samples, all 13 Vietnamese samples were identified as "Vietnamese" and the remaining 23 Chinese samples were identified as "not Vietnamese." Changing the order of applying the indicator variable ranges for the Korean and Vietnamese samples produced the same results, indicating that the two sets of ranges were well separated.

Setting the range of discriminant indicator variables aids in determining the authenticity of food, based on the content of intrinsic ingredients (such as metabolites and minerals) [28,37–39]. However, it is not easy to apply this method to discriminate between the origins of the same food. Therefore, it is meaningful that the geographical origin was correctly classified by setting several discriminant indicators and their ranges. Recently, FT-IR spectroscopy combined with statistical analysis has been actively applied to determine the authenticity, adulteration, and geographical origins of various foods. If the discriminant indicator variables and their ranges are set suitably, more effective and practical use of such results can be realized.

Table 6. Reclassification results for red pepper powder samples using the ranges of the indicator variables.

Applied Ranges	Origin	Predicted Results			Accuracy (%)
		Predicted	Not Predicted	Total	
Korean	Korean	50	0	50	100
	Not Korean	0	36	36	100
Vietnamese	Vietnamese	13	0	13	100
	Not Vietnamese	13	0	13	100

3.3. Blind Tests

To evaluate the applicability of the developed statistical discrimination method and the discriminant indicator variables to new samples, a blind test was performed on 19 new red pepper powder samples (9 Korean, 5 Chinese, and 5 Vietnamese), which were not used in the previous statistical analyses. The geographical origins were correctly classified for all the 19 blind red pepper powder samples using the established canonical discriminant functions (Table 7).

Table 7. Classification of the geographical origins of the blind samples using the established canonical discriminant functions.

Sample No.	Function 1	Function 2	Origin	Predicted	Probability
1	−1.075	−2.451	Korean	Korean	0.998
2	0.995	−0.257	Korean	Korean	0.991
3	2.643	−1.205	Korean	Korean	1
4	2.118	−0.816	Korean	Korean	1
5	1.979	−1.449	Korean	Korean	1
6	1.018	−0.926	Korean	Korean	0.999
7	1.49	−0.881	Korean	Korean	1
8	1.66	0.506	Korean	Korean	0.963
9	0.757	−0.917	Korean	Korean	0.998
10	−0.434	2.797	Chinese	Chinese	1
11	−1.023	2.405	Chinese	Chinese	1
12	−1.402	0.23	Chinese	Chinese	0.946
13	−1.825	7.135	Chinese	Chinese	1
14	−0.862	4.166	Chinese	Chinese	1
15	−5.508	1.616	Vietnamese	Vietnamese	1
16	−4.216	−0.415	Vietnamese	Vietnamese	1
17	−4.529	1.186	Vietnamese	Vietnamese	0.977
18	−5.688	0.018	Vietnamese	Vietnamese	1
19	−7.532	−0.289	Vietnamese	Vietnamese	1

Table 8 presents the classification results of comparing the values of the indicator peak variables obtained from the second-derivative FT-IR spectra of the blind samples with the discriminant ranges for the Korean and Vietnamese red peppers. When the ranges of the indicator variables for the Korean red pepper powder were applied, nine blind samples were correctly identified as "Korean", and the remaining 10 blind samples were classified as "not Korean". When applying the ranges of indicator variables for Vietnamese pepper to 10 blind "not Korean" samples, five samples were correctly identified as "Vietnamese". As in the canonical discriminant analysis, the other five samples that were classified as neither Korean nor Vietnamese can be assumed to be Chinese red pepper powder samples. These results indicate that the indicator ranges can be conveniently used to classify the geographical origins of new red pepper powder samples, even if they are established using a limited number of samples.

Table 8. Classification of the blind samples based on their geographical origins using the ranges of the indicator variables for Korean and Vietnamese samples.

Sample No.	P8	P12	P14	P17	Origin	Predicted
1	2.703	0.804	1.097	0.624	Korean	Korean
2	2.637	0.749	1.081	0.794	Korean	Korean
3	2.379	0.549	0.96	0.858	Korean	Korean
4	1.328	0.414	0.712	0.808	Korean	Korean
5	0.955	0.268	0.577	0.753	Korean	Korean
6	1.562	0.465	0.709	0.729	Korean	Korean
7	1.623	0.522	0.772	0.782	Korean	Korean
8	2.393	0.781	0.912	0.873	Korean	Korean
9	1.903	0.637	0.753	0.752	Korean	Korean
10	4.197	1.525	1.437	0.935	Chinese	(Chinese) *
11	4.544	1.515	1.284	0.841	Chinese	(Chinese) *
12	3.684	1.287	1.4	0.759	Chinese	(Chinese) *
13	7.074	2.778	2.22	1.117	Chinese	(Chinese) *
14	5.164	1.858	1.534	0.921	Chinese	(Chinese) *
15	4.362	1.777	1.648	0.69	Vietnamese	Vietnamese
16	5.107	1.818	1.564	0.604	Vietnamese	Vietnamese
17	5.32	1.919	1.608	0.694	Vietnamese	Vietnamese
18	5.67	2.1	1.824	0.617	Vietnamese	Vietnamese
19	6.971	2.468	1.955	0.559	Vietnamese	Vietnamese

* Samples were classified as neither Korean nor Vietnamese.

4. Conclusions

In this study, we investigated the feasibility of second-derivative FT-IR spectroscopy, combined with multivariate statistical analysis, to discriminate red pepper samples from Korea, China, and Vietnam, based on their geographical origins. Canonical discriminant functions for classifying Asian red pepper powders based on geographical origins were derived from the discriminant analysis, and the discriminating capability of the functions was verified by 100% correct reclassification of the origins of the powder samples used in the analysis. The results of the blind test to classify new red pepper powder samples according to geographical origins confirmed that the derived discriminant functions could correctly classify all new test samples. Although the classification method using the canonical discrimination functions is highly accurate, it requires the statistical data and program used to create the functions to discriminate the origins of new samples. To compensate for these limitations and simply determine the geographical origin without a special statistical program, four indicator variables with large differences in values according to their origins were selected from the variables used in the statistical analysis, and their origin-specific ranges were set. These indicator ranges were successfully used to correctly classify the geographical origins of all statistical samples and blind samples. Although applied to a limited number of samples, the use of the ranges of discriminant indicator variables provides a simple classification method for new samples. Further analyses of more red pepper powder samples, including samples from other countries, may enhance the capability and accuracy of the method of using both the canonical discriminant functions and the discriminant indicator variable ranges. In addition, the discriminant method that uses set ranges of the discriminant indicator variables may be useful in terms of experimental methodology; however, it can be expected to have more applications useful in fields that manage the traceability of foods.

In conclusion, the findings of this study indicate that the second-derivative FT-IR spectroscopy is a reliable, low-cost, and convenient analytical method for discriminating Asian red pepper powders according to their geographical origins.

Supplementary Materials: The following are available online at https://www.mdpi.com/article/10.3390/foods10051034/s1, Figure S1: Expanded (fructose hydrogen regions, normalized to the integral sum of all signals) ^{1}H NMR spectra of Asian red pepper powders at 600 MHz NMR. Table S1: Test of homogeneity of variance between the groups of Korean, Chinese, and Vietnamese red pepper powders using second derivative values of FT-IR spectra. Table S2: Pearson's Correlation Matrix.

Author Contributions: Conceptualization, M.K. and S.A.; methodology, H.S.C., B.H.K., and S.A.; software, M.K., J.H., D.L., and S.K.; validation, M.K., J.H., D.L., and S.A.; formal analysis, M.K. and J.H.; investigation, M.K., J.H., D.L., and S.K.; resources, H.S.C., Y.-H.C., and S.A.; data curation, J.H. and S.A.; writing—original draft preparation, M.K. and J.H.; writing—review and editing, J.H., B.H.K., and S.A.; visualization, M.K. and J.H.; supervision, S.A.; project administration, B.H.K. and S.A.; funding acquisition, Y.-H.C., H.S.C., and S.A. All authors have read and agreed to the published version of the manuscript.

Funding: This research was supported by the Chung-Ang University Research Scholarship in 2019 and a grant 17162MFDS065 from Ministry of Food and Drug Safety in 2019.

Conflicts of Interest: The authors declare no conflict of interest.

References

1. Shin, K.O.; MoriTani, T. Alterations of Autonomic Nervous Activity and Energy Metabolism by Capsaicin Ingestion during Aerobic Exercise in Healthy Men. *J. Nutr. Sci. Vitaminol.* **2007**, *53*, 124–132. [CrossRef] [PubMed]
2. Martinez, S.; Lopez, M.; Gonzalez-Raurich, M.; Bernardo Alvarez, A. The effects of ripening stage and processing systems on vitamin C content in sweet peppers (*Capsicum annuum* L.). *Int. J. Food Sci. Nutr.* **2005**, *56*, 45–51. [CrossRef] [PubMed]
3. Srinivasan, K. Biological Activities of Red Pepper (*Capsicum annuum*) and Its Pungent Principle Capsaicin: A Review. *Crit. Rev. Food Sci Nutr.* **2016**, *56*, 1488–1500. [CrossRef] [PubMed]
4. Sanatombi, K.; Rajkumari, S. Effect of Processing on Quality of Pepper: A Review. *Food Rev. Int.* **2019**, *36*, 626–643. [CrossRef]
5. Schupp, A.; Gillespie, J. Consumer Attitudes Toward Potential Country-of-Origin Labeling of Fresh or Frozen Beef. *J. Food Distrib. Res.* **2001**, *32*, 1–11. [CrossRef]
6. Hong, E.; Lee, S.Y.; Jeong, J.Y.; Park, J.M.; Kim, B.H.; Kwon, K.; Chun, H.S. Modern analytical methods for the detection of food fraud and adulteration by food category. *J. Sci Food Agric.* **2017**, *97*, 3877–3896. [CrossRef]
7. Choi, J.-Y.; Bang, K.-H.; Han, K.-Y.; Noh, B.-S. Discrimination Analysis of the Geographical Origin of Foods. *Korean J. Food Sci. Technol.* **2012**, *44*, 503–525. [CrossRef]
8. Kim, E.; Lee, S.; Baek, D.; Park, S.; Lee, S.; Ryu, T.; Lee, S.; Kang, H.; Kwon, O.; Kil, M.; et al. A comparison of the nutrient composition and statistical profile in red pepper fruits (*Capsicums annuum* L.) based on genetic and environmental factors. *Appl. Biol. Chem.* **2019**, *62*, 48. [CrossRef]
9. Willenberg, I.; Parma, A.; Bonte, A.; Matthaus, B. Development of Chemometric Models Based on a LC-qToF-MS Approach to Verify the Geographic Origin of Virgin Olive Oil. *Foods* **2021**, *10*, 479. [CrossRef] [PubMed]
10. Zhou, Y.; Kim, S.Y.; Lee, J.S.; Shin, B.K.; Seo, J.A.; Kim, Y.S.; Lee, D.Y.; Choi, H.K. Discrimination of the Geographical Origin of Soybeans Using NMR-Based Metabolomics. *Foods* **2021**, *10*, 435. [CrossRef] [PubMed]
11. Becerra-Martinez, E.; Florentino-Ramos, E.; Perez-Hernandez, N.; Gerardo Zepeda-Vallejo, L.; Villa-Ruano, N.; Velazquez-Ponce, M.; Garcia-Mendoza, F.; Banuelos-Hernandez, A.E. 1H NMR-based metabolomic fingerprinting to determine metabolite levels in serrano peppers (*Capsicum annum* L.) grown in two different regions. *Food Res. Int.* **2017**, *102*, 163–170. [CrossRef]
12. Ma, Y.; Tian, J.; Wang, X.; Huang, C.; Tian, M.; Wei, A. Fatty Acid Profiling and Chemometric Analyses for Zanthoxylum Pericarps from Different Geographic Origin and Genotype. *Foods* **2020**, *9*, 1676. [CrossRef] [PubMed]
13. Li, A.; Yat, Y.W.; Yap, W.K.; Lim, C.W.; Chan, S.H. Discriminating authentic Nostoc flagelliforme from its counterfeits by applying alternative ED-XRF and FTIR techniques. *Food Chem.* **2011**, *129*, 528–532. [CrossRef] [PubMed]
14. Ruoff, K.; Luginbühl, W.; Künzli, R.; Iglesias, M.T.; Bogdanov, S.; Bosset, J.O.; Ohe, K.v.d.; Ohe, W.v.d.; Amado, R. Authentication of the Botanical and Geographical Origin of Honey by Mid-Infrared Spectroscopy. *J. Agric. Food Chem.* **2006**, *54*, 6873–6880. [CrossRef] [PubMed]
15. Lee, B.J.; Zhou, Y.; Lee, J.S.; Shin, B.K.; Seo, J.A.; Lee, D.; Kim, Y.S.; Choi, H.K. Discrimination and prediction of the origin of Chinese and Korean soybeans using Fourier transform infrared spectrometry (FT-IR) with multivariate statistical analysis. *PLoS ONE* **2018**, *13*, e0196315. [CrossRef]
16. Ordoudi, S.A.; de los Mozos Pascual, M.; Tsimidou, M.Z. On the quality control of traded saffron by means of transmission Fourier-transform mid-infrared (FT-MIR) spectroscopy and chemometrics. *Food Chem.* **2014**, *150*, 414–421. [CrossRef]
17. Tapp, H.S.; Defernez, M.; Kemsley, E.K. FTIR Spectroscopy and Multivariate Analysis Can Distinguish the Geographic Origin of Extra Virgin Olive Oils. *J. Agric. food Chem.* **2003**, *51*, 6110–6115. [CrossRef] [PubMed]
18. Hennessy, S.; Downey, G.; O'Donnell, C. Multivariate analysis of attenuated total reflection-Fourier transform infrared spectroscopic data to confirm the origin of honeys. *Appl. Spectrosc.* **2008**, *62*, 1115–1123. [CrossRef] [PubMed]
19. Valand, R.; Tanna, S.; Lawson, G.; Bengtstrom, L. A review of Fourier Transform Infrared (FTIR) spectroscopy used in food adulteration and authenticity investigations. *Food Addit Contam Part. A Chem Anal. Control. Expo. Risk Assess.* **2020**, *37*, 19–38. [CrossRef]

20. Giannetti, V.; Mariani, M.B.; Marini, F.; Torrelli, P.; Biancolillo, A. Grappa and Italian spirits: Multi-platform investigation based on GC–MS, MIR and NIR spectroscopies for the authentication of the Geographical Indication. *Microchem. J.* **2020**, *157*, 104896. [CrossRef]
21. Ordoudi, S.A.; Papapostolou, M.; Kokkini, S.; Tsimidou, M.Z. Diagnostic Potential of FT-IR Fingerprinting in Botanical Origin Evaluation of Laurus nobilis L. Essential Oil is Supported by GC-FID-MS Data. *Molecules* **2020**, *25*, 583. [CrossRef]
22. Maione, C.; Barbosa, F.; Barbosa, R.M. Predicting the botanical and geographical origin of honey with multivariate data analysis and machine learning techniques: A review. *Comput. Electron. Agric.* **2019**, *157*, 436–446. [CrossRef]
23. Lee, D.; Kim, M.; Kim, B.H.; Ahn, S. Identification of the Geographical Origin of Asian Red Pepper (*Capsicum annuum* L.) Powders Using 1H NMR Spectroscopy. *Bull. Korean Chem. Soc.* **2020**, *41*, 317–322. [CrossRef]
24. Yin, X.; Xu, X.; Zhang, Q.; Xu, J. Rapid Determination of the Geographical Origin of Chinese Red Peppers (*Zanthoxylum Bungeanum* Maxim.) Based on Sensory Characteristics and Chemometric Techniques. *Molecules* **2018**, *23*, 1001. [CrossRef] [PubMed]
25. Zhang, J.; Yang, R.; Chen, R.; Li, Y.C.; Peng, Y.; Wen, X. Geographical origin discrimination of pepper (*Capsicum annuum* L.) based on multi-elemental concentrations combined with chemometrics. *Food Sci Biotechnol.* **2019**, *28*, 1627–1635. [CrossRef]
26. Song, B.Y.; Ryu, J.S.; Shin, H.S.; Lee, K.S. Determination of the source of bioavailable Sr using $^{87}Sr/^{86}Sr$ tracers: A case study of hot pepper and rice. *J. Agric. Food Chem.* **2014**, *62*, 9232–9238. [CrossRef] [PubMed]
27. Kohler, A.; Bertrand, D.; Martens, H.; Hannesson, K.; Kirschner, C.; Ofstad, R. Multivariate image analysis of a set of FTIR microspectroscopy images of aged bovine muscle tissue combining image and design information. *Anal. Bioanal. Chem.* **2007**, *389*, 1143–1153. [CrossRef] [PubMed]
28. Park, S.M.; Yu, H.Y.; Chun, H.S.; Kim, B.H.; Ahn, S. A Second Derivative Fourier-Transform Infrared Spectroscopy Method to Discriminate Perilla Oil Authenticity. *J. Oleo. Sci.* **2019**, *68*, 389–398. [CrossRef] [PubMed]
29. Rieppo, L.; Saarakkala, S.; Narhi, T.; Helminen, H.J.; Jurvelin, J.S.; Rieppo, J. Application of second derivative spectroscopy for increasing molecular specificity of Fourier transform infrared spectroscopic imaging of articular cartilage. *Osteoarthr. Cartil.* **2012**, *20*, 451–459. [CrossRef]
30. Mossoba, M.M.; Milosevic, V.; Milosevic, M.; Kramer, J.K.; Azizian, H. Determination of total trans fats and oils by infrared spectroscopy for regulatory compliance. *Anal. Bioanal. Chem.* **2007**, *389*, 87–92. [CrossRef] [PubMed]
31. Kim, H.Y. Analysis of variance (ANOVA) comparing means of more than two groups. *Restor. Dent. Endod.* **2014**, *39*, 74–77. [CrossRef] [PubMed]
32. Rencher, A.C. Interpretation of Canonical Discriminant Functions, Canonical Variates, and Principal Components. *Am. Stat.* **1992**, *46*, 217–225. [CrossRef]
33. Gautam, R.; Vanga, S.; Ariese, F.; Umapathy, S. Review of multidimensional data processing approaches for Raman and infrared spectroscopy. *EPJ Tech. Instrum.* **2015**, *2*, 8. [CrossRef]
34. Baker, M.J.; Trevisan, J.; Bassan, P.; Bhargava, R.; Butler, H.J.; Dorling, K.M.; Fielden, P.R.; Fogarty, S.W.; Fullwood, N.J.; Heys, K.A.; et al. Using Fourier transform IR spectroscopy to analyze biological materials. *Nat. Protoc.* **2014**, *9*, 1771–1791. [CrossRef]
35. Zhou, W.; Guo, P.; Chen, J.; Lei, Y. A rapid analytical method for the quantitative determination of the sugar in acarbose fermentation by infrared spectroscopy and chemometrics. *Spectrochim. Acta A Mol. Biomol Spectrosc.* **2020**, *240*, 118571. [CrossRef] [PubMed]
36. Jeon, H.; Kim, I.-H.; Lee, C.; Choi, H.-D.; Kim, B.H.; Akoh, C.C. Discrimination of Origin of Sesame Oils Using Fatty Acid and Lignan Profiles in Combination with Canonical Discriminant Analysis. *J. Am. Oil Chem. Soc.* **2012**, *90*, 337–347. [CrossRef]
37. Kim, J.; Jin, G.; Lee, Y.; Chun, H.S.; Ahn, S.; Kim, B.H. Combined Analysis of Stable Isotope, ^1H NMR, and Fatty Acid To Verify Sesame Oil Authenticity. *J. Agric. Food Chem.* **2015**, *63*, 8955–8965. [CrossRef]
38. Lee, J.E.; Hwang, J.; Choi, E.; Shin, M.J.; Chun, H.S.; Ahn, S.; Kim, B.H. Rubidium analysis as a possible approach for discriminating between Korean and Chinese perilla seeds distributed in Korea. *Food Chem.* **2020**, *312*, 126067. [CrossRef] [PubMed]
39. Kim, J.; Yang, S.; Jo, C.; Choi, J.; Kwon, K.; Ahn, S.; Sook Chun, H.; Hee Kim, B. Comparison of Carbon Stable Isotope and Fatty Acid Analyses for the Authentication of Perilla Oil. *Eur. J. Lipid Sci. Technol.* **2018**, *120*, 1700480. [CrossRef]
40. Christy, A.A.; Egeberg, P.K. Quantitative determination of saturated and unsaturated fatty acids in edible oils by infrared spectroscopy and chemometrics. *Chemom. Intell. Lab. Syst.* **2006**, *82*, 130–136. [CrossRef]
41. Guillen, M.D.; Cabo, a.N. Infrared Spectroscopy in the Study of Edible Oils and Fats. *J. Sci Food Agric.* **1997**, *75*, 1–11. [CrossRef]
42. Lammers, K.; Arbuckle-Keil, G.; Dighton, J. FT-IR study of the changes in carbohydrate chemistry of three New Jersey pine barrens leaf litters during simulated control burning. *Soil Biol. Biochem.* **2009**, *41*, 340–347. [CrossRef]
43. Smith, B.C. An IR Spectral Interpretation Potpourri: Carbohydrates and Alkynes. *Spectroscopy* **2017**, *32*, 18–24. Available online: http://www.spectroscopyonline.com/ir-spectral-interpretation-potpourri-carbohydrates-and-alkynes (accessed on 22 April 2021).
44. Wang, L.; Yang, Q.; Zhao, H. Sub-regional identification of peanuts from Shandong Province of China based on Fourier transform infrared (FT-IR) spectroscopy. *Food Control.* **2021**, *124*, 107879. [CrossRef]
45. Wiercigroch, E.; Szafraniec, E.; Czamara, K.; Pacia, M.Z.; Majzner, K.; Kochan, K.; Kaczor, A.; Baranska, M.; Malek, K. Raman and infrared spectroscopy of carbohydrates: A review. *Spectrochim. Acta A Mol. Biomol. Spectrosc.* **2017**, *185*, 317–335. [CrossRef] [PubMed]
46. Coenen, K.; Gallucci, F.; Mezari, B.; Hensen, E.; van Sint Annaland, M. An in-situ IR study on the adsorption of CO_2 and H_2O on hydrotalcites. *J. CO_2 Util.* **2018**, *24*, 228–239. [CrossRef]

47. Huberty, C.J. Discriminant Analysis_Carl J Huberty. *Rev. Educ. Res.* **1975**, *45*, 543–598. [CrossRef]
48. Schellenberg, A.; Chmielus, S.; Schlicht, C.; Camin, F.; Perini, M.; Bontempo, L.; Heinrich, K.; Kelly, S.D.; Rossmann, A.; Thomas, F.; et al. Multielement stable isotope ratios (H, C, N, S) of honey from different European regions. *Food Chem.* **2010**, *121*, 770–777. [CrossRef]
49. Shi, X.; Li, J.; Wang, S.; Zhang, L.; Qiu, L.; Han, T.; Wang, Q.; Chang, S.K.; Guo, S. Flavor characteristic analysis of soymilk prepared by different soybean cultivars and establishment of evaluation method of soybean cultivars suitable for soymilk processing. *Food Chem.* **2015**, *185*, 422–429. [CrossRef] [PubMed]
50. Ananthan, R.; Subhash, K.; Longvah, T. Capsaicinoids, amino acid and fatty acid profiles in different fruit components of the world hottest Naga king chilli (*Capsicum chinense* Jacq). *Food Chem.* **2018**, *238*, 51–57. [CrossRef] [PubMed]

Article

Rapid Authentication of Potato Chip Oil by Vibrational Spectroscopy Combined with Pattern Recognition Analysis

Siyu Yao [1], Didem Peren Aykas [1,2] and Luis Rodriguez-Saona [1,*]

[1] Department of Food Science and Technology, The Ohio State University, 110 Parker Food Science and Technology Building, 2015 Fyffe Road, Columbus, OH 43210, USA; yao.806@osu.edu (S.Y.); aykas.1@osu.edu (D.P.A.)
[2] Department of Food Engineering, Faculty of Engineering, Adnan Menderes University, Aydin 09100, Turkey
* Correspondence: rodriguez-saona.1@osu.edu; Tel.: +1-614-292-3339

Abstract: The objective of this study was to develop a rapid technique to authenticate potato chip frying oils using vibrational spectroscopy signatures in combination with pattern recognition analysis. Potato chip samples (n = 118) were collected from local grocery stores, and the oil was extracted by a hydraulic press and characterized by fatty acid profile determined by gas chromatography equipped with a flame ionization detector (GC-FID). Spectral data was collected by a handheld Raman system (1064 nm) and a miniature near-infrared (NIR) sensor, further being analyzed by SIMCA (Soft Independent Model of Class Analogies) and PLSR (Partial Least Square Regression) to develop classification algorithms and predict the fatty acid profile. Supervised classification by SIMCA predicted the samples with a 100% sensitivity based on the validation data. The PLSR showed a strong correlation (Rval > 0.97) and a low standard error of prediction (SEP = 1.08–3.55%) for palmitic acid, oleic acid, and linoleic acid. 11% of potato chips (n = 13) indicated a single oil in the label with a mislabeling problem. Our data supported that the new generation of portable vibrational spectroscopy devices provided an effective tool for rapid in-situ identification of oil type of potato chips in the market and for surveillance of accurate labeling of the products.

Keywords: rapid authentication; handheld Raman; NIR; fatty acid profile; oil qualification

Citation: Yao, S.; Aykas, D.P.; Rodriguez-Saona, L. Rapid Authentication of Potato Chip Oil by Vibrational Spectroscopy Combined with Pattern Recognition Analysis. *Foods* **2021**, *10*, 42. https://dx.doi.org/10.3390/foods10010042

Received: 20 November 2020
Accepted: 21 December 2020
Published: 25 December 2020

Publisher's Note: MDPI stays neutral with regard to jurisdictional claims in published maps and institutional affiliations.

Copyright: © 2020 by the authors. Licensee MDPI, Basel, Switzerland. This article is an open access article distributed under the terms and conditions of the Creative Commons Attribution (CC BY) license (https://creativecommons.org/licenses/by/4.0/).

1. Introduction

The potato chip was invented 167 years ago and has been the most popular snack food in America for more than 50 years [1,2]. Oil represents between 25% and 35% weight of the potato chip, serving as the heat transfer agent and providing the flavor and texture of the product [3]. As reported by researchers, the main precursors of volatile compounds in potato chips are polyunsaturated fatty acids in the frying oil [4–6]. The non-heterogeneous oil distribution during the frying contributes to the surface color of potato chips [7]. The common types of oil utilized in potato chip manufacturing are corn, sunflower (mid-oleic and high-oleic varieties), canola, high-oleic (HO) safflower, and cottonseed oils [8].

As the trend toward wellness keeps gaining strength, the selection of oils can add value as healthier alternatives. For example, systematic studies suggested that consuming foods rich in monounsaturated or polyunsaturated fat positively affected blood glucose control, compared with consuming saturated fat or dietary carbohydrate, and may help to prevent metabolic diseases [9,10]. Accordingly, numerous potato chip manufacturers are selecting oils with high-oleic traits to meet buyer healthier preferences. However, adulteration of high-price oils is a prevalent source of economically-motivated fraud [11]. Canola, soybean, and palm oils become common adulterants for high price oils like sunflower oil, which has a higher content of unsaturated fatty acid [12]. Therefore, there is an urgent need for authentication and prevention of adulteration for the sake of consumers and honest companies.

Accurate and appropriate analytical methods are required to identify the oil type based on their components [13,14]. Traditionally, fatty acid methyl esters (FAMEs) are analyzed by gas chromatography with flame ionization detector (GC-FID) to determine oil types based on the fatty acid composition, and Iodine value (IV) is utilized to classify oils according to their degree of unsaturation [15–17]. However, these conventional methods are labor-intensive, time-consuming, high-priced, require the use of harmful reagents and generate hazardous waste [18]. Hence, it is necessary to develop technologies that can provide real-time screening and in-field applications to authenticate the oil used in potato chip manufacturing. Vibrational spectroscopy (near-infrared (NIR), mid-infrared (mid-IR) and Raman) are rapid methods to offer a high throughput, simple, sensitive and robust technique for establishing reliable authentication for raw materials, based on their specific signature profiles coupled with pattern recognition techniques [19].

Raman spectroscopy (50–8000 cm^{-1}) is based on the inelastic scattering of monochromatic light [20,21]. When the sample interacts with the monochromatic laser, in addition to the relatively more pronounced elastic scattering effect in the mode of Rayleigh scattering, an inelastic scattering can arise which results in new photon emissions with different frequencies or a shift from that of the excitation light. This scattering is called Raman scattering, whereby Raman shifts are directly related to the vibrational states of a molecule structure [22]. Near-infrared (NIR) spectroscopy (800–2500 nm) is based on molecular overtone and combination vibrations in the region of the electromagnetic spectrum [23]. For a molecule to be Raman active, the polarizability of the molecule needs to be changed through incident radiation and a center of symmetry is required, while for NIR activity to be dominant, the dipole moment of the molecule has to be changed and, thus, the molecule ought not have a center of symmetry. Therefore, usually the molecules which are Raman active are not IR active and vice versa [24].

Meanwhile, advancement in semiconductors has allowed the miniaturization of the components such as solid-state lasers, wavelength selectors, and detectors leading to commercially accessible and affordable portable, handheld, compact, and micro-vibrational spectroscopy devices in the industry [19]. These portable/handheld spectrometers have the tremendous potential capability to move from the confines of the comparatively steady and controlled laboratory setting to the potentially more dynamic and complex environments at- or in-line, at points of vulnerability along complicated food supply chains [25].

However, limited information is reported in the literature regarding the rapid authentication of oils used in manufacturing potato chips using vibrational spectroscopy. Aykas et al. [8] evaluated a portable MIR in conjunction with pattern recognition analysis to develop classification methods for the authentication of potato chip oils. Nonetheless, the measurement process needs heating for preventing oil solidification, which limits the in-field application. Baeten et al. [26] assessed the oil and fat classification by Raman spectroscopy (1064 nm) by using principal component analysis (PCA) that was applied to 138 samples from 21 different sources and reported that stepwise linear discriminant analysis can classify oils based on their unique monounsaturated and polyunsaturated composition. Dong et al. [27] established a predictive model of the fatty acid composition of vegetable oil based on least squares support vector machines (LS-SVM), by Raman (785 nm) spectral data. McDowell et al. [24] built calibration models with four different multivariate classifiers (soft independent modeling of class analogy (SIMCA), linear discriminant analysis—k-nearest neighbor (LDA-KNN), partial least squares—discriminant analysis (PLS-DA), and linear discriminant analysis—support vector machine (LDA-SVM)) based on either FT-IR and Raman spectral fingerprints to detect the oil addition in cold-pressed rapeseed, achieving high sensitivity of 86% and 93%, respectively, when refined sunflower oil is the adulterant. These studies have shown the potential capabilities of vibrational spectroscopy to detect vegetable oil adulteration. However, they do not show sufficient ability to classify based on different types of vegetable oils, and they have not applied the analysis to oil expelled from the real food matrix. Moreover, most have been developed using a limited number of oil types, limiting their application as a reliable method to detect oil adulteration of food products in the market [28].

The objective of this study was to develop a rapid detection method to identify the type of oil used in the manufacturing of potato chips and to predict the fatty acid profile of the oil based on the unique Raman and NIR spectral patterns.

2. Materials and Methods

A total of 118 potato chip samples, including 102 samples for generating the training models and 16 samples serving as an independent external validation set, were collected from local grocery stores in Columbus, OH. The potato chips (~10 g) were pressed to expel oil (~3 g) by a manual hydraulic press (3851 Benchtop Laboratory Manual Press, Carver, Inc., Wabash, IN, USA). The crushed potato chips filled a stainless-steel cylinder container. The oil was expelled by applying pressure on the cylinder to 10,000 psi for 1 min. Oil is collected and stored at 3 °C in the glass vials for further analysis. Six different reference vegetable oils, including corn, canola, sunflower (high-oleic and mid-oleic), peanut, and cottonseed oils, were collected from online vendors and local stores.

2.1. Reference Method

The reference method for obtaining the fatty acid profile is based on a fatty acid methyl ester (FAME) procedure with modification [29]. Methyl ester structures were produced by dissolving 100 µL oil sample with 1 mL of hexane into a 2 mL centrifuge tube, and the mixture was vortexed. Then 20 µL 2 N potassium hydroxide in methanol was added to the centrifuge tube and vortexed for 1 min. The upper hexane part was transferred to a new 2 mL centrifuge tube with one pinch of sodium sulfate anhydrous and centrifuged at 4000 rpm for 10 min. After that, 500 µL supernatant was transferred into a 2 mL GC glass vial and mixed with 700 µL hexane thoroughly for further analysis. FAME profile analysis was done in duplicate for all samples by an Agilent 6890 arrangement (Agilent Technologies, Inc., Santa Clara, CA, USA) gas chromatograph (GC) equipped with a flame ionization detector (FID), an Agilent 7693 autosampler (Agilent Technologies, Inc., Santa Clara, CA, USA), and a tray. The fatty acids were separated by utilizing an HP-88 60 m × 0.25 mm × 0.2 mm (Agilent 112-8867, Agilent Technologies, Inc., Santa Clara, CA, USA)) GC column and utilizing helium as the carrier gas. The injection volume was 0.1 µL, with a split ratio of 60.3: 1. The inlet and detector temperatures were 250 °C. The oven temperature was set at 120 °C held for 1 min as the initial, then at 175 °C (10 °C/min) held for 10 min, then at 210 °C (4 °C/min) held for 4 min and finally at 230 °C (4 °C/min) held for 4.75 min. Based on the reference standards (Supelco® 37 Component FAME Mix, Sigma Aldrich, Inc., St. Louis, MO, USA), through the comparison of each peak's retention times, fatty acids were identified [28]. All the samples (n = 118) were analyzed by GC-FID, and if the fatty acid composition of the sample matched with the profiles of reference oils or literature values, this sample was identified as being fried by the corresponding single oil source; otherwise, it was determined as being fried using oil mixtures.

2.2. Spectral Data Acquisition

2.2.1. Raman Spectral Data Acquisition

A handheld Raman instrument, Progeny™ (Rigaku Analytical Devices, Inc., Wilmington, MA, USA) equipped with a 1064 nm excitation laser (Figure 1a), was used to analyze the oil (at least 500 µL required) in the transparent glass vial obtained from the pressing process. The Raman device equipped with a thermoelectrically cooled InGaAs 512-pixel detector operated at 8 cm^{-1} spectral resolution with a spectral range of 200–2500 cm^{-1} [30]. The laser power and exposure time were set at 230 mW and 3 s, respectively, with 15 averages to maximize the signal-to-noise ratio. A background was collected after the spectrum was collected for each sample. The spectra were collected in duplicate for all samples (n = 118).

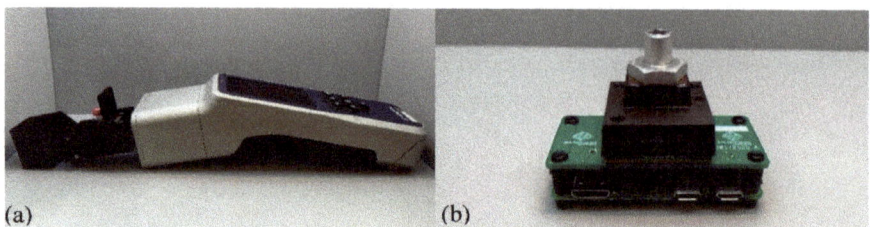

Figure 1. Potato chip oil spectrum acquisition by (**a**) using a handheld Raman instrument equipped with a 1064 nm excitation laser and by (**b**) using a compact Fourier Transform Near-Infrared (FT-NIR) spectral sensor.

2.2.2. NIR Spectral Data Acquisition

The NIR spectral data was collected by the NeoSpectra Micro (Si-Ware Systems, Inc., Cairo, Egypt), which is a compact Fourier Transform Near-Infrared (FT-NIR) spectral sensor with a single uncooled InGaAs photodetector utilizing a single-chip Michelson interferometer with monolithic opto-electro-mechanical structure based on Fourier Transform Infrared (FT-IR) technology [31]. A 100 µL oil aliquot was deposited onto the sensor of the spectrometer and the oil was covered with a reflectance accessory, NIRA Liquids Sample Accessory (Perkin Emerto, Inc., Llantrisant, Pontyclun, UK) to perform the measurement as shown in Figure 1b. An oil spectrum was collected in duplicate for all samples ($n = 118$) over the range of 1350–2552 nm in absorbance mode and a resolution of 25 cm^{-1}. To get the best reproducibility and signal-to-noise ratio, the scanning time was set to 20 s.

2.3. Multivariate Data Analysis

The spectral data were analyzed by multivariate statistical analysis software, Pirouette® (version 4.5, Infometrix, Inc., Bothell, WA, USA). Raman spectral data was transformed by normalization (sample 2-norm), where each data value was divided by the sample's maximum value for SIMCA and PLSR analysis. NIR spectra were pre-processed by autoscaling to correct for different scaling and units, and transformed by Savitsky–Golay second derivative (15 points with second-order polynomial filter) and Smoothing (to help reduce baseline noise) in the NIR SIMCA analysis. In the Raman and NIR PLSR analysis, mean-centering was utilized as the preprocessing method to alleviate "micro" but not "macro" multicollinearity [32].

The classification algorithm of potato chip oil was generated using the SIMCA method, a supervised classification method that clusters oil samples with common Raman or near-infrared spectral features and distinguishes them into their vegetable oil sources with different profiles based on principal component analysis (PCA) [33]. Samples were divided into training (83 single vegetable oil source samples verified by their FAME assignments) and external validation (16 samples, single oil and oil mixture samples) sets. The training set is utilized to "teach" the system about the Raman and NIR spectral features of each population (class) to determine whether discrimination differences are present, which is accomplished by providing the model with the class assignments based on GC-FID data. External validation of the SIMCA model's performance was evaluated by an unseen independent dataset (16 samples) using the trained model, generating an unbiased estimation of the resembling model deployment for predictions in a real situation and determining if these potato chip oils match their "market" labels [34]. SIMCA model performance was evaluated in terms of misclassifications (percentage of samples correctly assigned to their original groups), class projections, discriminating power (most significant regions or wavenumbers for class separations), and interclass distances (ICD) describing the similarity or dissimilarity of the different classes quantitatively, it being accepted generally that samples can be well-differentiated when ICD > 3 [35].

PLSR is a quantitative technique for generating quantitative training predictive models through combining characteristics from multiple linear regression and PCA [30]. Raman and NIR spectra of all 102 samples (single oil source and oil mixture samples) were correlated with their fatty acid profile for developing PLSR predictive models. The performance of PLSR models for predicting fatty acid compositions were evaluated using leave-one-out as the internal cross-validation and an unseen independent dataset (16 samples) was set to validate the models externally. PLSR model performance was evaluated in terms of correlation coefficients (R^2), residual analysis, outlier diagnostics, leverage, standard error cross-validation (SECV), and the standard error of prediction (SEP) [8]. If the leverage and/or studentized residual is high for a sample, this sample has a high possibility to be an outlier, and it was excluded from the model [28].

3. Results and Discussion

3.1. Characterization of Potato Chip Frying Oil (Fatty Acid Composition and Spectral Analysis)

To generate a training model for identifying the oil type used in the manufacturing, all the oils extracted from the potato chip samples were profiled based on the GC-FID method. Among all the samples (n = 102), based on their fatty acid profiles, 19 samples were identified as being fried using oil mixtures, while 83 samples were manufactured with a single vegetable oil source. The fatty acid compositions (C16:0, C18:0, C18:1 n-9, C18:2 n-6 and C18:3 n-3) of samples with single oil source were summarized in Table 1, including corn oil (n = 22), canola oil (n = 8), mid-oleic sunflower oil (n = 14), high-oleic sunflower oil (I) (n = 14), high-oleic sunflower oil (II) (n = 16), peanut oil (n = 4), and cottonseed oil (n = 5). Overall, cottonseed oil (17.6–21.8%) and corn oil (8.4–14.1%) showed the highest content of palmitic acid, while HO sunflower (I) oil (82.0–87.1%) showed the highest content of oleic acid, and cottonseed (57.0–59.1%) and corn oil (54.5–58.5%) showed the highest content of linoleic acid (Table 1).

Table 1. Fatty acid composition summary of oil from potato chip samples and oil references using gas chromatograph flame ionization detector (GC-FID) method.

		Corn	Canola	HO SUN [a] (I)	HO SUN [a] (II)	MO SUN [b]	Peanut	Cottonseed
Palmitic (%) C16:0	Range	8.4–14.1	2.9–4.7	2.4–5.2	2.5–4.9	3.5–5.9	3.0–5.0	17.6–21.8
	Mean	11	3.9	4.2	3.9	4.5	4.2	20
	SD	1.4	0.7	0.8	0.7	0.7	0.9	1.6
	Reference oils	9.6	3.9	2.8	—	3.4	8.1	16.8
Stearic (%) C18:0	Range	1.6–2.3	1.9–2.1	2.9–3.8	1.7–4.3	2.1–4.2	2.5–3.3	2.6–3.2
	Mean	1.9	2	3.4	3.2	3.5	2.9	2.9
	SD	0.2	0.1	0.3	0.9	0.8	0.4	0.3
	Reference oils	1.8	1.9	2.6	—	3.5	3.1	2.9
Oleic (%) C18:1 n-9	Range	28.3–32.3	66.6–68.7	82.0–87.1	70.9–78.9	64.1–69.9	75.6–81.4	18.9–20.1
	Mean	30.5	67.6	83.9	74.3	67.6	78.5	19.2
	SD	0.9	0.7	1.7	2.5	1.6	2.6	0.5
	Reference oils	30	66.5	84.3	—	66.6	66.6	20.6
Linoleic (%) C18:2 n-6	Range	54.5–58.5	18.4–19.5	6.7–10.4	14–21.9	22.5–27.6	11.4–15.4	57.0–59.1
	Mean	55.7	19.1	8.4	17.7	24.3	13.8	57.9
	SD	1	0.4	1.2	2.5	1.3	1.7	0.9
	Reference oils	57.6	19.4	10.3	—	26	25.9	59.4
Linolenic (%) C18:3 n-3	Range	0.6–1.0	6.2–9.1	0.0–0.8	0.2–2.4	0.1–1.4	0–0.8	0.0–0.2
	Mean	0.9	7.5	0.2	0.8	0.4	0.4	0.2
	SD	0.1	1.1	0.2	0.8	0.3	0.4	0.1
	Reference oils	1.1	8.4	0	—	0.5	0.4	0.2

[a] HO SUN: a high-range oleic, above 70% monounsaturated sunflower oil; [b] MO SUN: a mid-range oleic, around 65% monounsaturated sunflower oil.

To confirm the accuracy of oil type identification, fatty acid composition of oil from potato chip samples was compared with reference oils (Table 1) and literature values. The fatty acid profiles of corn, canola, high-oleic sunflower (I), mid-oleic sunflower and cottonseed oils were in agreement with our reference oils, and those reported by Caballero et al., Aykas et al., and Dubois et al. [8,36–38]. The peanut oil extracted from potato chip had a higher content of oleic acid (75.6–81.4%) and a lower content of linoleic acid (11.4–15.4%), compared to the values these researchers reported (around 52.1% and 32.9%, respectively). However, their fatty acid values fell into the fatty acid composition range (oleic acid: 52.8–82.2%, linoleic acid: 2.9–27.1%) found by Worthington et al. [39] for most cultivated peanuts. The discrepancies in the fatty acid composition under the same oil source can be related to differences in geographic origin and variety of seed-cultivars, and in seed and oil processes [40]. Interestingly, in the case of sunflower oil, three different fatty acid profiles (MO sunflower, HO sunflower (I) and HO sunflower (II)) were found. Stability of oil is directly related to its degree of unsaturation, and HO sunflower oils, which have over 70% oleic acid, are more stable than their counterparts with higher content of polyunsaturated fatty acids, linoleic and linolenic acids, fulfilling a better performance in the heating tolerance for a longer fry life [41–43]. The varieties of HO sunflower (I) oil containing over 80% oleic acid and HO sunflower (II) oil containing from 70% to 80% oleic acid can come from genetic selection, naturally occurring variation and trough mutagenesis [44].

Figure 2a showed the overlapped Raman spectra of seven different potato chip oils (cottonseed, peanut, HO sunflower (I), HO sunflower (II), MO sunflower, canola, and corn oils) and the corresponding band assignments. The band existing at 1745 cm^{-1} was the stretching vibration of ester bond carbonyl. The band at 1659 cm^{-1} was associated with C=C stretching (cis-R-HC=CH-R) from polyunsaturated fatty acids, while the band at 1263 cm^{-1} corresponds with in-plane =C-H deformation in an unconjugated cis (C=C), which was associated with monounsaturated fatty acids. The band at 1443 cm^{-1} was associated with CH$_2$ scissoring deformation (δCH$_2$), and the band at 1300 cm^{-1} was related to in-phase methylene twisting motion. The band at 1080 cm^{-1} was associated with the stretching vibration of the methylene chain skeleton [28,45]. As can be seen in Figure 2a, the signal to noise ratio was excellent across the spectral region and the Raman spectra patterns for these oils were similar to each other, but they appear to show an obviously different intensity on the bands of stretching (cis-R-HC=CH-R), shear bending (-CH$_2$) and stretching (=C-H). An increase in the stretching (cis-R-HC=CH-R) and stretching (=C-H) bands intensity is correlated to the increase of unsaturated fatty acids weight percentage in oils [46], while the ratio of stretching (cis-R-HC=CH-R) to shear bending(-CH$_2$) is inversely correlated with the content of saturated fatty acid [47].

Figure 2b showed the characteristic NIR absorption spectra of the seven different potato chip oil examples demonstrating the close similarity in their spectral characteristics. The peaks in NIR spectra were much broader compared with Raman. Briefly, characteristic NIR absorbance bands arise in four regions in the spectrum. Region A (1350–1490 nm) results from the combinations of C-H stretching and bending. Region B (1640–1885 nm) corresponds with the first overtone of the C-H stretching vibration of several chemical groups (methyl, methylene and ethylene groups). Furthermore, Region C (2050–2230 nm) is related to the C–H vibration of cis-unsaturation, and the intensity increasing in this region reflects the increase in the degree of total unsaturation. The two peaks in attributed fat could be observed clearly in the region D (2310–2350 nm), which represents the characteristic of the combination of C-H stretching vibration and other vibrational modes [48–50].

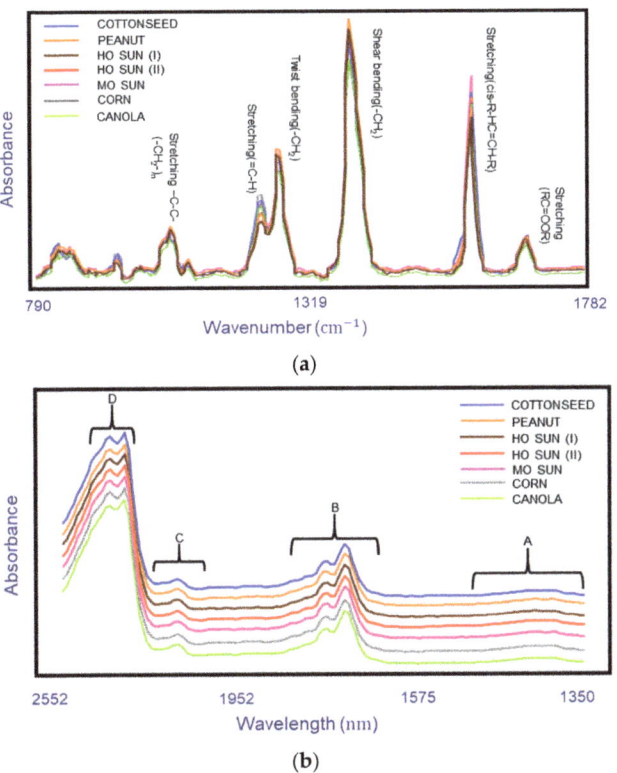

Figure 2. (**a**) Raman spectra and band assignments of some vegetable oil examples collected by a handheld Raman instrument equipped with a 1064 nm excitation laser. (**b**) Near-infrared (NIR) spectra and important absorbance regions of vegetable oils collected by a miniature NIR sensor.

3.2. Pattern Recognition Modeling for Raman and NIR Spectroscopy

The Raman and NIR spectral data were analyzed using SIMCA for the classification and rapid authentication of different frying potato chip oils based on the FAME profile. The class projection plot of the training SIMCA model generated with Raman spectral data (Figure 3a) showed distinctive clustering patterns and seven well-defined groups for different sole source oils in the three-dimensional (3D) environment. The interclass distances (ICD) shown in Table 2a describes the similarity or dissimilarity of the different classes quantitatively, ranging from 0.9 (MO SUN and HO SUN(II)) to 10.1 (HO SUN(I) and Corn) and it is generally accepted that samples can be differentiated when ICD > 3 [51]. Most of the classes, such as HO SUN(I) and MO SUN, HO SUN(I) and Canola Oil, HO SUN(I) and Corn oil, etc., are significantly differentiated between each other (ICD > 3), while some classes HO SUN(I) and HO SUN(II), HO SUN(I) and Peanut, MO SUN and Canola, MO SUN and HO SUN(II), HO SUN(II) and Peanut, and Corn and Cottonseed gave ICD < 3 because of the limited compositional difference among them [8]. In order to discriminate between the classes and minimize the overfitting problem, five principal components were employed to explain 99% of the variance. The discriminating power graph (Figure 3c) in the SIMCA model defines the variables (wavenumbers) mainly responsible for the potato chip oil classification [33], which can be representative of specific chemical structures. The band centered at 1659 cm^{-1} was associated with (cis-R-HC=CH-R) from polyunsaturated fatty acids, which has the most significant influence on classifying the samples. The band at 1443 cm^{-1} corresponded to the CH$_2$ scissoring deformation, and bands at 1252 and 1267 cm^{-1} were related to stretching(=C-H), monounsaturated fatty acids.

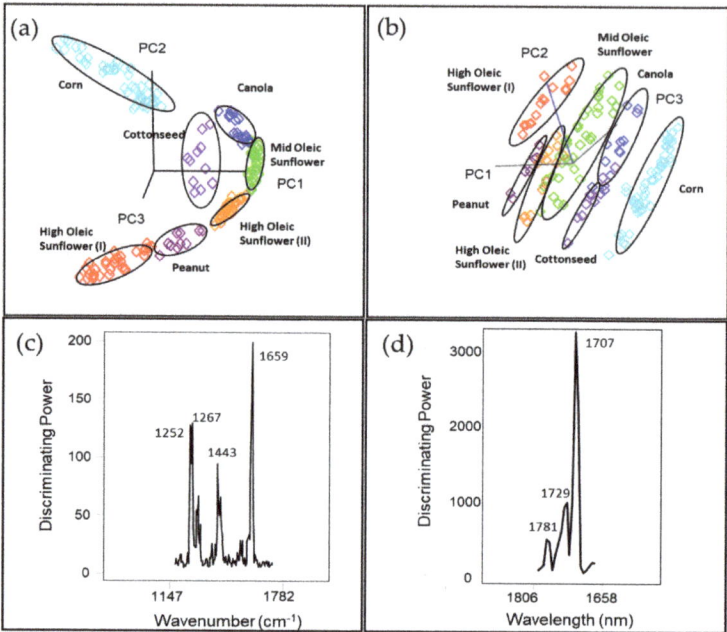

Figure 3. Soft Independent Model of Class Analogies (SIMCA) 3D projection plots of spectral data for potato chip oil samples collected by (**a**) a handheld Raman spectrometer (1064 nm) and (**b**) a miniature near Infrared (NIR) sensor. (**c**) Discriminating plots of Raman and (**d**) NIR SIMCA models showing bands and regions responsible for class separation.

Table 2. Interclass distance between 7 types of potato chip frying oil based on the SIMCA training model generated by (**a**) the Raman spectral data collected in the 790–1782 cm^{-1} region and (**b**) NIR spectral data collected in the 1350–2552 nm region.

Groups	HO SUN (I)	MO SUN	Canola	HO SUN (II)	Peanut	Corn	Cottonseed
(a)							
HO SUN (I)	0.0						
MO SUN	3.6	0.0					
Canola	7.1	1.5	0.0				
HO SUN (II)	2.0	0.9	3.3	0.0			
Peanut	1.3	3.1	6.5	1.3	0.0		
Corn	10.1	4.5	3.2	5.8	9.7	0.0	
Cottonseed	7.2	3.7	3.0	3.8	7.0	2.6	0.0
(b)							
HO SUN (I)	0.0						
MO SUN	3.8	0.0					
Canola	44.8	11.8	0.0				
HO SUN (II)	6.2	2.9	25.3	0.0			
Peanut	8.7	10.5	34.7	7.2	0.0		
Corn	13.0	5.5	15.5	14.5	39.0	0.0	
Cottonseed	40.2	14.5	13.9	26.1	36.1	12.0	0.0

The class projection of the SIMCA model generated by NIR spectral data (Figure 3b) showed similar grouping patterns obtained from Raman, but it improved class separation with larger interclass distances, yielding well-defined clusters using three to five principal components. There was no misclassification under the cross-validation and the interclass distances (Table 2b) among different classes of samples varying between 2.9 and 44.8. The highest ICD (44.8) was between HO SUN(I) and Canola oil, while there was only one group of classes that had an ICD < 3, which was between MO SUN and HO SUN(II). The SIMCA discriminating plot (Figure 3d) illustrated that the clustering of different potato chip oils was explained by the wavelength associated with 1707, 1729, and 1781 nm, corresponding to the first overtone of the C-H stretching vibration of several chemical groups (methyl, methylene, and ethylene groups).

The predictive accuracy of SIMCA training models generated by the Raman and NIR spectral data was evaluated using an independent external validation set that included 16 commercial potato chip samples. Among them, only six samples were labeled with a single oil as their frying sources, including cottonseed, sunflower and expeller-pressed sunflower oils, and the remaining (n = 10) were labeled as having one or more type of oils. Figure 4a,b showed the Raman and NIR SIMCA 3D projection for the external validation set, respectively. Figure 4c summarized their label information, GC-FID analysis results, and Raman and NIR SIMCA predictions. Our GC-FID results showed that 12 out of 16 samples were manufactured with one type of vegetable oil, including corn, HO SUN(I), HO SUN(II) and cottonseed oils. Our Raman and NIR SIMCA predictions were consistent with the GC-FID assignments for all these 12 samples. Besides, 4 samples (E, F, I and M) were identified as having oil mixtures (two or more types of oils) based on their fatty acid profiles. SIMCA predictions of both Raman and NIR instruments indicated Sample I fried with oil mixtures and the GC-FID assignment confirmed; however, its label falsely indicated it as containing only sunflower oil. GC-FID assignment showed that sample E contained canola oil as its main component and at least one other type of oil. In the Raman and NIR SIMCA projection plots, Sample E was clustered close to canola and MO SUN classes in the 3D environment. Sample E was predicted as a mixture accurately in the NIR SIMCA prediction. However, due to the small interclass distance (1.5) between canola and MO SUN classes in the Raman SIMCA model, the oil from sample E was predicted as canola oil instead of the oil mixture in the Raman SIMCA prediction. The oil from Sample F was identified as a mixture based on its GC-FID result. In the Raman SIMCA projection plot, this oil mixture was clustered very close to the canola group, which led to the false prediction as canola oil. On the other hand, the NIR SIMCA model accurately predicted sample F as the oil mixture, though this sample was clustered close to the canola group in the NIR projection. Our results demonstrated some compositional similarities between canola oil and sample E and F. Sample M was also identified as an oil mixture based on GC-FID, and it was projected in the space closed to canola and corn clusters in the Raman and NIR projection plots. Raman and NIR SIMCA models both predicted sample M accurately as an oil mixture.

Sensitivity determined the ability of the classification model to identify the sole oil type of potato chips, while specificity evaluated the capability of our model to discriminate the oil mixture from the sole oil types [28]. The predictive performance statistics of the NIR SIMCA model showed 100% sensitivity ($n_{true\ positive}$ = 12, $n_{false\ negative}$ = 0) and 100% specificity ($n_{false\ positive}$ = 0, $n_{true\ negative}$ = 4) (Table 3) in classifying the independent samples, matching the results obtained from the GC-FID method. The Raman SIMCA model showed 100% sensitivity ($n_{true\ positive}$ = 12, $n_{false\ negative}$ = 0) and 50% specificity ($n_{false\ positive}$ = 2, $n_{true\ negative}$ =2) (Table 3) since Sample E and F which are oil mixtures based on the GC-FID results falsely predicted as samples using a sole oil source.

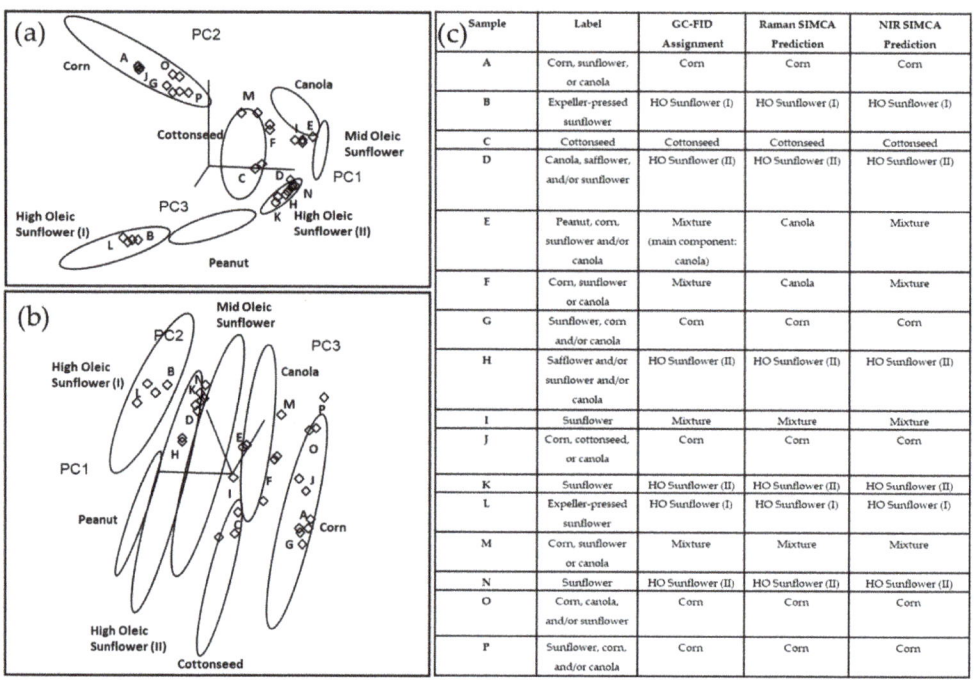

Figure 4. (a) Raman and (b) NIR SIMCA projection for the external validation set (*n* = 16). (c) Information summary of manufacture's label claims, GC-FID assignments, Raman SIMCA predictions and NIR SIMCA predictions for the external validation set.

Table 3. Specificity and sensitivity values of SIMCA models obtained from the handheld Raman (1064 nm) and the miniature NIR spectral data.

Model Types	Sensitivity (%)	Specificity (%)
Raman	100	50
NIR	100	100

Similar to our research using the Raman approach, Yang et al. [50] used linear discriminant analysis (LDA) and canonical variate analysis (CVA) to discriminate corn oil, peanut oil, canola oil, safflower oil, etc., resulting in about 94% classification accuracy with their FT-Raman equipment. In addition, Velioglu et al. [52] differentiated seven vegetable oils successfully using principal component analysis (PCA) by Raman spectroscopic barcode. Similar to our NIR approach, Yang et al. [50] differentiated oils using LDA and CVA with 93% accuracy with their FT-NIR equipment, and Bewig et al. [53] discriminated vegetable oils successfully by NIR reflectance spectroscopy. Based on these previous studies, we explored a novel strategy to apply supervised pattern recognition that allows us to predict the oil type in the further application, and we also analyzed the ability of our model to predict the oil mixture. In addition, to our best knowledge, our study is the first in the literature to apply Raman and NIR to the potato chip (food matrix) oil authentication.

Our model generated by using the Raman and NIR spectra coupled with pattern recognition analysis has adequate ability to rapidly (~1 min for Raman, ~20 sec for NIR) authenticate the mislabeling problem in potato chip products and be a potentially useful tool to perform in-situ screening of potato chip oil types in the market.

3.3. PLSR Models for Raman and NIR Spectroscopy

Saturated (SFA), monounsaturated (MUFA) and polyunsaturated fatty acids (PUFA), palmitic acid (C16:0), oleic acid (C18:1 n-9) and linoleic acid (C18:2 n-6), respectively, were predominant in vegetable oils and their contents are related to oil and product stability and quality [33]. Therefore, it is crucial to monitor the major fatty acid content in oil during potato chip manufacturing and storage [54]. The quantitative models, partial least squares regression (PLSR) models, were developed using the handheld Raman (1064 nm) and NIR spectral data based on the reference value of fatty acid composition (Figure 5). The performance statistics of PLSR models generated using a calibration (n = 102) and external validation (n = 16) data set are summarized in Table 4. The number of samples and the range in calibration models are not all the same because of the outlier exclusion [28]. Six factors were chosen to generate all the FTIR and Raman calibration models based on the standard error of cross-validation (leave-one-out) result, achieving the best quality of the models and avoiding the risk of overfitting at the same time [55].

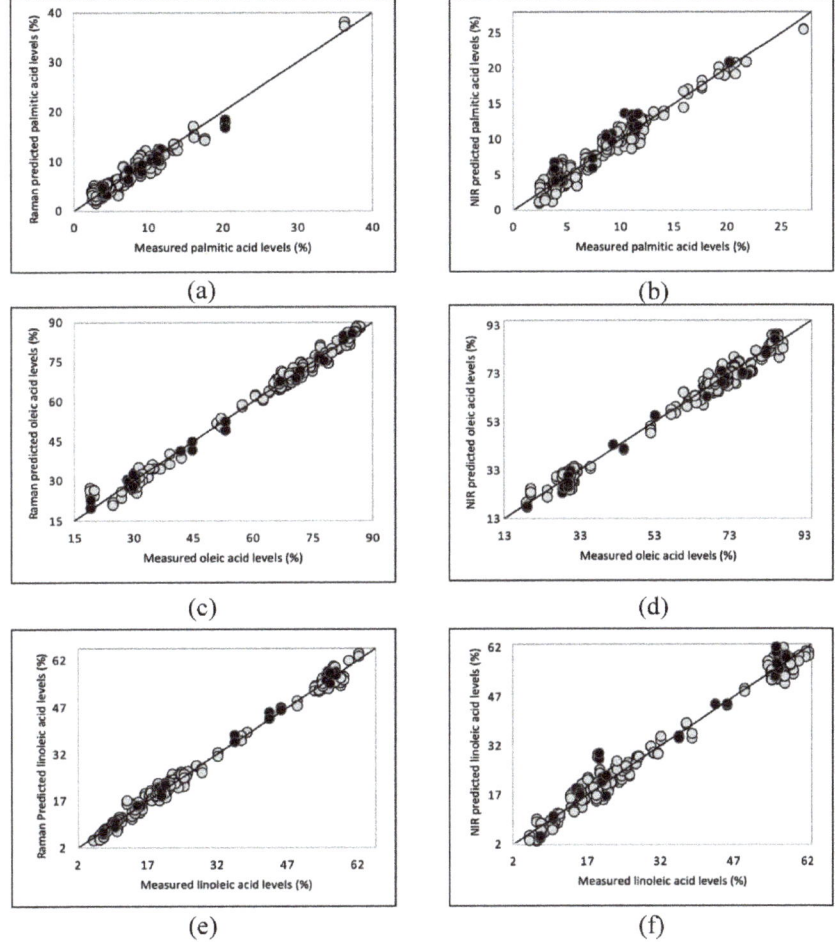

Figure 5. PLSR calibration and validation plots for main fatty acids, palmitic acid (**a**,**b**), oleic acid (**c**,**d**), and linoleic acid (**e**,**f**) in potato chip samples utilizing Raman and NIR data respectively.

Table 4. The performance statistics of Partial Least Square Regression (PLSR) models developed using a training ($n = 102$) and an external validation ($n = 16$) data set based on Raman and NIR spectral data for estimating palmitic, oleic, linoleic acid composition in potato chip samples.

Approach	Fatty Acid	Training Model					External Validation Model			
		Range	N [a]	Factor	SECV [b]	Rcal	Range	N [c]	SEP [d]	Rval
Raman	Palmitic (%)	2.4–36.3	101	6	1.08	0.98	3.7–20.3	16	1.08	0.97
	Oleic (%)	18.9–86.9	102	6	2.26	1	19.1–84.9	16	1.84	1
	Linoleic (%)	5.3–62.4	102	6	1.48	1	7.5–57.5	16	1.31	1
NIR	Palmitic (%)	2.5–27.2	94	6	1.06	0.98	3.7–20.3	16	1.60	0.97
	Oleic (%)	18.9–86.9	95	6	2.61	0.99	19.1–84.9	16	2.87	0.99
	Linoleic (%)	5.3–62.4	101	6	2.47	0.99	7.5–57.5	16	3.55	0.99

[a] Sample number in the training models. [b] Standard error of cross validation. [c] Sample number in the external validation models. [d] Standard error of prediction.

Our PLSR models showed a strong correlation (Rcal > 0.98 and Rval > 0.97) in predicting palmitic, oleic, and linoleic acid content in potato chip oils. The standard error of prediction (SEP) values, ranging from 1.08%–1.84% for the three predominant fatty acids in Raman validation models and ranging from 1.60%–3.55% for NIR external validation models, are similar to the standard error of cross validation (leave-one-out) values in each calibration model which demonstrate the robustness of the models. Overall, the Raman regression models demonstrated superior performance than those generated by the NIR sensor, especially for linoleic acid. The correlation coefficient of validation and SEP for linoleic acid obtained from the Raman model was 1 and 1.31%, respectively. In contrast, the NIR model gave a Rval of 0.99 and a SEP of 3.55%. Our handheld Raman units demonstrated better performance for the prediction of the main fatty acids composition (higher Rcal and Rval) than the study reported by Dong and others (2013) for vegetable oils using a portable Raman spectrometer with a shorter wavelength laser (785 nm) coupled with least squares support vector machines [27]. Meanwhile, our NIR models showed superior performance on higher Rval in predicting oleic and linoleic acids when compared with the past research on oils conducted by Casale et al. [56] and lower SEP in predicting oleic acid compared with the study reported by Sato [57] using their benchtop NIR units.

4. Conclusions

This study showed that a handheld Raman device with 1064 nm excitation laser and a miniature NIR sensor allowed for rapid authentication of the oil type used in potato chip manufacturing. Based on the result of GC-FID analysis, a total of 83 (~70%) potato chip samples were identified as having been manufactured with a single oil, including corn oil (19%), canola oil (7%), mid-oleic sunflower oil (12%), high-oleic sunflower (I) (12%), high-oleic sunflower (II) (14%), peanut oil (3%) and cottonseed oil (4%). Combining the pattern recognition analysis, potato chip oils were successfully clustered into their corresponding oil type used in frying and our external validation set demonstrated a 100% accuracy for identifying single oils by using Raman and NIR models. Interestingly, pattern recognition predictions showed that 11% of potato chips ($n = 13$) that indicated a single oil in the label were mislabeled, which was corroborated by GC-FID analysis. In addition, the same spectra allowed the prediction of the major fatty acid composition (palmitic acid, oleic acid and linoleic acid) with strong correlation (Rval > 0.97) and low standard error of prediction. The performance of the PLSR models obtained from the handheld Raman device were superior to models from portable Raman units in other studies and comparable to results from benchtop infrared systems. The handheld Raman spectrometer and miniature NIR sensor can provide applicable tools to perform the rapid authentication of potato chip oil type and in-situ determination of their main fatty acid composition in the market.

Author Contributions: S.Y.; methodology, formal analysis, data curation, validation, writing-original draft preparation. D.P.A.; validation, resources, writing-review and editing. L.R.-S.; conceptualization, methodology, data curation, validation, writing-review and editing. All authors have read and agreed to the published version of the manuscript.

Funding: This research received no external funding.

Acknowledgments: The authors would like to thank ADM, Incorporated for providing NuSun oil.

Conflicts of Interest: Authors declare that they have no conflict of interest. This article does not contain any studies with human or animal subjects.

References

1. Northern Plains Potato Growers Association Potato Fun facts. Available online: http://nppga.org/consumers/funfacts.php (accessed on 10 September 2020).
2. Grand View Research U.S. Potato Chips Market Size, Share & Trends Analysis Report By Flavor (Flavored, Plain/Salted), By Distribution Channel (Supermarket, Convenience Stores), And Segment Forecasts, 2018–2025. Available online: https://www.grandviewresearch.com/industry-analysis/us-potato-chips-market (accessed on 10 June 2020).
3. Process Sensors Corporation. Potato Chip Moisture and Oil. Available online: https://www.processsensors.com/industries/food/potato-chip-moisture-oil (accessed on 26 September 2020).
4. Maarse, H. Vegetables. In *Volatile Compounds in Foods and Beverages*; Marcel Dekke, Inc.: New York, NY, USA, 1991; pp. 223–281.
5. Maga, J.A. Potato flavor. *Food Rev. Int.* **1994**, *10*, 1–48. [CrossRef]
6. Martin, F.L.; Ames, J.M. Comparison of flavor compounds of potato chips fried in palmolein and silicone fluid. *JAOCS J. Am. Oil Chem. Soc.* **2001**, *78*, 863–866. [CrossRef]
7. Pedreschi, F.; Mery, D.; Marique, T. Quality Evaluation and Control of Potato Chips and French Fries. *Comput. Vis. Technol. Food Qual. Eval.* **2008**, 545–566. [CrossRef]
8. Aykas, D.P.; Rodriguez-Saona, L.E. Assessing potato chip oil quality using a portable infrared spectrometer combined with pattern recognition analysis. *Anal. Methods* **2016**, *8*, 731–741. [CrossRef]
9. Imamura, F.; Micha, R.; Wu, J.H.Y.; de Oliveira Otto, M.C.; Otite, F.O.; Abioye, A.I.; Mozaffarian, D. Effects of Saturated Fat, Polyunsaturated Fat, Monounsaturated Fat, and Carbohydrate on Glucose-Insulin Homeostasis: A Systematic Review and Meta-analysis of Randomised Controlled Feeding Trials. *PLoS Med.* **2016**, *13*, 1–18. [CrossRef] [PubMed]
10. Pimpin, L.; Wu, J.H.Y.; Haskelberg, H.; Del Gobbo, L.; Mozaffarian, D. Is butter back? A systematic review and meta-analysis of butter consumption and risk of cardiovascular disease, diabetes, and total mortality. *PLoS ONE* **2016**, *11*, 1–18. [CrossRef] [PubMed]
11. Moore, J.C.; Spink, J.; Lipp, M. Development and Application of a Database of Food Ingredient Fraud and Economically Motivated Adulteration from 1980 to 2010. *J. Food Sci.* **2012**, *77*. [CrossRef]
12. Jee, M. *Oils and Fats Authentication*; Blackwell Publishing Ltd.: Oxford, UK, 2002; pp. 1–24.
13. Salimon, J.; Omar, T.A.; Salih, N. An accurate and reliable method for identification and quantification of fatty acids and trans fatty acids in food fats samples using gas chromatography. *Arab. J. Chem.* **2017**, *10*, S1875–S1882. [CrossRef]
14. Giacomelli, L.M.; Mattea, M.; Ceballos, C.D. Analysis and characterization of edible oils by chemometric methods. *JAOCS J. Am. Oil Chem. Soc.* **2006**, *83*, 303–308. [CrossRef]
15. Kang, J.X.; Wang, J. A simplified method for analysis of polyunsaturated fatty acids. *BMC Biochem.* **2005**, *6*, 4–7. [CrossRef]
16. Kyriakidis, N.B.; Katsiloulis, T. Calculation of iodine value from measurements of fatty acid methyl esters of some oils: Comparison with the relevant American Oil Chemists Society method. *J. Am. Oil Chem. Soc.* **2000**, *77*, 1235–1238. [CrossRef]
17. Dijkstra, A.J. *Vegetable Oils: Composition and Analysis*, 1st ed.; Elsevier Ltd.: Amsterdam, The Netherlands, 2015; ISBN 9780123849533.
18. Nunes, C.A. Vibrational spectroscopy and chemometrics to assess authenticity, adulteration and intrinsic quality parameters of edible oils and fats. *Food Res. Int.* **2014**, *60*, 255–261. [CrossRef]
19. Rodriguez-Saona, L.E.; Giusti, M.M.; Shotts, M. *Advances in Infrared Spectroscopy for Food Authenticity Testing*; Elsevier Ltd.: Amsterdam, The Netherlands, 2016; ISBN 9780081002209.
20. Larkin, P. General Outline and Strategies for IR and Raman Spectral Interpretation. *Infrared Raman Spectrosc.* **2011**, 117–133. [CrossRef]
21. Jennifer Line Raman Spectroscopy and the Analysis of Gemstones. Available online: https://www.sas.upenn.edu/~{}lineje/ramanspectroscopy.html (accessed on 14 December 2020).
22. Cherry, S.R.; Sorenson, J.A.; Phelps, M.E. *Physics in Nuclear Medicine*; Elsevier Ltd.: Amsterdam, The Netherlands, 2012; ISBN 9781416051985.
23. Stuart, B.H. *Infrared Spectroscopy: Fundamentals and Applications*; Analytical Techniques in the Sciences; John Wiley & Sons, Ltd.: Chichester, UK, 2004; ISBN 9780470011140.
24. McDowell, D.; Osorio, M.T.; Elliott, C.T.; Koidis, A. Detection of Refined Sunflower and Rapeseed Oil Addition in Cold Pressed Rapeseed Oil Using Mid Infrared and Raman Spectroscopy. *Eur. J. Lipid Sci. Technol.* **2018**, *120*, 1–10. [CrossRef]

25. Ellis, D.I.; Muhamadali, H.; Haughey, S.A.; Elliott, C.T.; Goodacre, R. Point-and-shoot: Rapid quantitative detection methods for on-site food fraud analysis-moving out of the laboratory and into the food supply chain. *Anal. Methods* **2015**, *7*, 9401–9414. [CrossRef]
26. Baeten, V.; Hourant, P.; Morales, M.T.; Aparicio, R. Oil and Fat Classification by FT-Raman Spectroscopy. *J. Agric. Food Chem.* **1998**, *46*, 2638–2646. [CrossRef]
27. Dong, W.; Zhang, Y.; Zhang, B.; Wang, X. Rapid prediction of fatty acid composition of vegetable oil by Raman spectroscopy coupled with least squares support vector machines. *J. Raman Spectrosc.* **2013**, *44*, 1739–1745. [CrossRef]
28. Aykas, D.P.; Karaman, A.D.; Keser, B.; Rodriguez-Saona, L. Non-targeted authentication approach for extra virgin olive oil. *Foods* **2020**, *9*, 221. [CrossRef]
29. Ichihara, K.N.; Shibahara, A.; Yamamoto, K.; Nakayama, T. An improved method for rapid analysis of the fatty acids of glycerolipids. *Lipids* **1996**, *31*, 535–539. [CrossRef]
30. Akpolat, H.; Barineau, M.; Jackson, K.A.; Akpolat, M.Z.; Francis, D.M.; Chen, Y.J.; Rodriguez-Saona, L.E. High-throughput phenotyping approach for screening major carotenoids of tomato by handheld raman spectroscopy using chemometric methods. *Sensors* **2020**, *20*, 3723. [CrossRef]
31. Aykas, D.P.; Ball, C.; Sia, A.; Zhu, K.; Shotts, M.-L.; Schmenk, A.; Rodriguez-Saona, L. In-Situ Screening of Soybean Quality with a Novel Handheld Near-Infrared Sensor. *Sensors* **2020**, *20*, 6283. [CrossRef]
32. Iacobucci, D.; Schneider, M.J.; Popovich, D.L.; Bakamitsos, G.A. Mean centering helps alleviate "micro" but not "macro" multicollinearity. *Behav. Res. Methods* **2016**, *48*, 1308–1317. [CrossRef] [PubMed]
33. Duckworth, J. Mathematical Data Preprocessing. In *Near-Infrared Spectroscopy in Agriculture*; American Society of Agronomy, Inc.: Madison, WI, USA, 2004; pp. 115–132.
34. Aykas, D.P.; Shotts, M.-L.; Rodriguez-Saona, L.E. Authentication of commercial honeys based on Raman fingerprinting and pattern recognition analysis. *Food Control* **2020**, *117*, 107346. [CrossRef]
35. Massart, D.L.; Vandeginste, B.G.M.; Deming, S.M.; Michotte, Y.; Kaufman, L. *Data Handling in Science and Technology*; Elsevier Ltd.: Amsterdam, The Netherlands, 2001; ISBN 9780444828538.
36. Caballero, B.; Trugo, L.C.; Finglas, P.M. *Encyclopedia of Food Sciences and Nutrition*; Elsevier Ltd.: Amsterdam, The Netherlands, 2003; ISBN 012227055X.
37. Dubois, V.; Breton, S.; Linder, M.; Fanni, J.; Parmentier, M. Fatty acid profiles of 80 vegetable oils with regard to their nutritional potential. *Eur. J. Lipid Sci. Technol.* **2007**, *109*, 710–732. [CrossRef]
38. Canolainfo.org. Classic and High-Oleic Canola Oils. Available online: https://www.canolacouncil.org/media/515008/classic_and_high_oleic_canola_oils.pdf (accessed on 8 November 2020).
39. Worthington, R.E.; Hammons, R.O.; Allison, J.R. Varietal Differences and Seasonal Effects on Fatty Acid Composition and Stability of Oil from 82 Peanut Genotypes. *J. Agric. Food Chem.* **1972**, *20*, 729–730. [CrossRef]
40. Vingering, N.; Oseredczuk, M.; Du Chaffaut, L.; Ireland, J.; Ledoux, M. Fatty acid composition of commercial vegetable oils from the French market analysed using a long highly polar column. *OCL Ol. Corps Gras Lipides* **2010**, *17*, 185–192. [CrossRef]
41. Oklahoma State University Canola Oil Properties. Available online: https://extension.okstate.edu/fact-sheets/canola-oil-properties.html (accessed on 8 November 2020).
42. Lindsay Nelson, R.D.L.D. High Oleic Sunflower Oil: Long Name, Great Benefits. Available online: https://fitjoyfoods.com/blogs/life-of-joy/high-oleic-sunflower-oil-long-name-great-benefits (accessed on 9 November 2020).
43. USDA High Oleic Sunflower Oil. Available online: https://fdc.nal.usda.gov/fdc-app.html#/food-details/521139/nutrients (accessed on 9 November 2020).
44. Fernández-Martínez, J.M.; Pérez-Vich, B.; Velasco, L.; Domínguez, J. Breeding for specialty oil types in sunflower. *Helia* **2007**, *30*, 75–84. [CrossRef]
45. Huang, F.; Li, Y.; Guo, H.; Xu, J.; Chen, Z.; Zhang, J.; Wang, Y. Identification of waste cooking oil and vegetable oil via Raman spectroscopy. *J. Raman Spectrosc.* **2016**, *47*, 860–864. [CrossRef]
46. Zhang, X.F.; Zou, M.Q.; Qi, X.H.; Liu, F.; Zhang, C.; Yin, F. Quantitative detection of adulterated olive oil by Raman spectroscopy and chemometrics. *J. Raman Spectrosc.* **2011**, *42*, 1784–1788. [CrossRef]
47. Bailey, G.F.; Horvat, R.J. Raman spectroscopic analysis of the cis/trans isomer composition of edible vegetable oils. *J. Am. Oil Chem. Soc.* **1972**, *49*, 494–498. [CrossRef]
48. Hourant, P.; Baeten, V.; Morales, M.T.; Meurens, M.; Aparicio, R. Oil and fat classification by selected bands of near-infrared spectroscopy. *Appl. Spectrosc.* **2000**, *54*, 1168–1174. [CrossRef]
49. García Martín, J.F.; López Barrera MD, C.; Torres García, M.; Zhang, Q.A.; Álvarez Mateos, P. Determination of the acidity of waste cooking oils by near infrared spectroscopy. *Processes* **2019**, *7*, 304. [CrossRef]
50. Yang, H.; Irudayaraj, J.; Paradkar, M.M. Discriminant analysis of edible oils and fats by FTIR, FT-NIR and FT-Raman spectroscopy. *Food Chem.* **2005**, *93*, 25–32. [CrossRef]
51. Kvalheim, O.M.; Karstang, T.V. SIMCA-classification by means of disjoint cross validated principal components models. In *Multivariate Pattern Recognition in Chemometrics, Illustrated by Case Studies*; Elsevier Ltd.: Amsterdam, The Netherlands, 1992; pp. 209–248.

52. Velioglu, S.D.; Ercioglu, E.; Temiz, H.T.; Velioglu, H.M.; Topcu, A.; Boyaci, I.H. Raman Spectroscopic Barcode Use for Differentiation of Vegetable Oils and Determination of Their Major Fatty Acid Composition. *JAOCS J. Am. Oil Chem. Soc.* **2016**, *93*, 627–635. [CrossRef]
53. Bewig, K.M.; Clarke, A.D.; Roberts, C.; Unklesbay, N. Discriminant analysis of vegetable oils by near-infrared reflectance spectroscopy. *J. Am. Oil Chem. Soc.* **1994**, *71*, 195–200. [CrossRef]
54. Orsavova, J.; Misurcova, L.; Vavra Ambrozova, J.; Vicha, R.; Mlcek, J. Fatty acids composition of vegetable oils and its contribution to dietary energy intake and dependence of cardiovascular mortality on dietary intake of fatty acids. *Int. J. Mol. Sci.* **2015**, *16*, 12871–12890. [CrossRef]
55. Shotts, M.L.; Plans Pujolras, M.; Rossell, C.; Rodriguez-Saona, L. Authentication of indigenous flours (Quinoa, Amaranth and kañiwa) from the Andean region using a portable ATR-Infrared device in combination with pattern recognition analysis. *J. Cereal Sci.* **2018**, *82*, 65–72. [CrossRef]
56. Casale, M.; Oliveri, P.; Casolino, C.; Sinelli, N.; Zunin, P.; Armanino, C.; Forina, M.; Lanteri, S. Characterisation of PDO olive oil Chianti Classico by non-selective (UV-visible, NIR and MIR spectroscopy) and selective (fatty acid composition) analytical techniques. *Anal. Chim. Acta* **2012**, *712*, 56–63. [CrossRef]
57. Sato, T. Nondestructive measurements of lipid content and fatty acid composition in rapeseeds (Brassica napus L.) by near infrared spectroscopy. *Plant Prod. Sci.* **2008**, *11*, 146–150. [CrossRef]

Article

SLE Single-Step Purification and HPLC Isolation Method for Sterols and Triterpenic Dialcohols Analysis from Olive Oil

Manuel León-Camacho * and María del Carmen Pérez-Camino

Lipid Characterization and Quality Department, Instituto de la Grasa, Spanish National Research Council, 41013 Seville, Spain
* Correspondence: mleon@ig.csic.es; Tel.: +34-954-611-550

Abstract: The unsaponifiable fraction of oils and fats constitutes a very small fraction but it is an essential part of the healthy properties of some specific oils. It is a complex fraction formed by a large number of minor compounds and it is a source of information to characterize and authenticate the oil sample. Specially, the composition of sterols of any oil or fat is a distinctive feature of itself and, therefore, it has become a useful tool for detecting contaminants and adulterants in oils. A new supported liquid extraction (SLE) technique for the analysis and characterization of the unsaponifiable fraction of fats and oils is proposed. The SLE system includes, as a stationary phase, a combination of adsorbent materials which allow a highly purified unsaponifiable matter ready to be isolated by high performance liquid chromatography (HPLC) and quantified by gas chromatography (GC). This method ensures the removal of fatty acids, avoiding possible interferences and making the analysis of sterols and triterpenic dialcohols easier. The procedure uses a small sample size (0.2 g), reduces the volume of solvents and reagents, and reduces the handling of samples subjected to analytical control. All this is achieved without losing either precision—a relative standard deviation of each compound lower than the reference value ($\leq$16.4%)—or recovery, being for all compounds higher than 88.00%. Therefore, this new technique represents a significant economic and time saving in business control laboratories, a larger productivity and enhancement of working safety.

Keywords: sterols; olive oil; triterpenic dialcohols; supported liquid extraction; high performance liquid chromatography; gas chromatography

1. Introduction

The authentication of foodstuff has been developed according to market tendencies, and analytical methods have evolved to detect adulterations.

The unsaponifiable fraction of oils or fats constitutes a very small fraction but an essential part of oils' healthy properties. It is a complex fraction formed by a large number of minor compounds. These compounds rarely represent more than 2% of the oil composition and include many compounds of a different nature [1].

In addition, the unsaponifiable matter is the most important fraction of edible fats and oils from the point of view of the characterization and verification of their authenticity. Its different compounds are used as chemical descriptors for the authentication of these products. In particular, the composition of sterols of any oil or fat is a distinctive feature of itself and, therefore, it has become a useful tool for the detection of contaminants and adulterants of oils [2,3]. In the case of olive oils, the analytical methods to determine the content in several sterols and triterpenic dialcohols and their values are described in detail in the official regulations [4–8].

Conventional methods for the determination of a sterol fraction consist of several steps: saponification, liquid–liquid extraction of the unsaponifiable matter, isolation of the 4-desmethylsterols, by either thin layer chromatography (TLC) or high performance liquid chromatography (HPLC), and quantification by gas chromatography with a flame ionization detector. All these procedures take a long time and require huge amounts of solvents and excessive handling [5–7,9,10].

The most critical step in the unsaponifiable fraction analysis is its purification to obtain the different groups of compounds. Thus, solid-phase extraction techniques have been developed to purify the unsaponifiable fraction, using C18 [11] as well as a silica cartridge [12,13] as an absorbent. However, these cartridges do not separate the unsaponifiable matter properly.

On several occasions, gas chromatography analysis of certain fractions of compounds from a more or less complex matrix, such as the entire unsaponifiable fraction, presents some difficulties, primarily when these fractions have a high number of compounds [14]. Furthermore, the performance of chemical reactions or derivatizations in order to isolate a specific fraction may cause problems due to alterations.

In that case, the previous isolation of the fraction via liquid–liquid extraction, despite being the standardized and official process [6], is not a very suitable procedure as it is tedious and takes a long time to perform. In addition, it is necessary to use a large oil sample (between 5 and 20 g) and solvent volume for the extraction (in the order of 300 mL); if emulsions (which commonly appear) come into play, it may take extra time. Finally, it is essential to eliminate the whole solvent used in the extraction; then, a purification needs to be done. Thus, before gas chromatography analysis, isolations have to be performed using thin layer chromatography, open glass column [15], solid phase extraction (SPE) [16], or high-performance liquid chromatography (HPLC) [5,7,17]. To sum up, in order to obtain the unsaponifiable fraction of an oil or fat and purify it through a conventional procedure, it is necessary to use a large sample, a high volume of organic solvents and a long period of time. All this may cause losses and contamination.

In the particular case of using a previous isolation technique, such as HPLC, it might be off-line or on-line, with a liquid chromatograph being needed in both cases to carry out the procedure. In the former case, the selected fraction is collected, the mobile phase is removed, and the fraction is transferred to the gas chromatograph using some of the already known techniques [1]. When on-line coupling techniques are used, a more or less complex interphase is required [18,19] in order to allow the transition from the liquid state (high pressure) to the gas state (low pressure); moreover, interfaces are very different depending on whether the liquid chromatography is absorbent or distributive [20–22].

The isolation of target compounds from unsaponifiable fractions using the aforementioned techniques is a critical step, especially in the case of olive oils, where sterols and triterpenic dialcohols elute in the same chromatographic zones, leading to incorrect results. The previous step (isolation of the unsaponifiable matter from the saponifiable fraction) is also critical, and, with this aim, in 1973 Hadorn and Zürcher [23] published one of the first attempts at isolation using column systems with a mixture of adsorbents.

An alternative to the abovementioned isolation techniques, which is not widely known, is supported liquid extraction (SLE). Basically, it consists of a chemically inert support, highly purified, used as a stationary phase to retain the aqueous phase, formed by phyllosilicates. The water is very easily adsorbed onto the surface of the phyllosilicate particles. The main phyllosilicate used in this technique is the diatomaceous earth. Johnson et al. reported this extraction method for the first time in 1997, using a calcinated diatomaceous earth called Hydromatrix [24].

SLE techniques for isolation of the unsaponifiable fraction have been developed using cartridges of diatomaceous earth Agilent Chem Elut, 20 mL, unbuffered, followed by filtering through an anhydrous sodium sulphate and an isolation of the fractions via solid phase extraction, using cartridges of activated silica with potassium, and eluting them with solutions of different ratios of hexane/diethyl ether. Next, sterol and triterpenic dialcohol fractions were derivatized and analyzed by gas chromatography [25].

The aim of this work was to develop a new SLE technique for the analysis and characterization of the unsaponifiable fraction of fats and oils. A SLE method that includes a stationary phase combining different adsorbent materials is presented. In addition, different solvent mixtures from that used in the literature (diethyl ether) are assayed. Finally, the use of the HPLC technique notably enhances the purification of the unsaponifiable matter, avoiding—as opposed to the described methods in the literature—interference from other unsaponifiable compounds. It will reduce sample size, the volume of solvents and reagents, and the handling of samples subjected to analytical control. Without losing precision and recovery while saving time and resources, this new technique will turn provide higher productivity and enhance working safety for business control laboratories.

2. Materials and Methods

2.1. Reagents and Solutions

Ethyl acetate and n-hexane of LiChrosolv grade were supplied by Merck (Darmstadt, Germany). Ethanol (96% vol.) and diethyl ether of analytical grade were supplied by VWR (Leuven, Belgium). Potassium hydroxide (85%) pellets and anhydrous sodium sulphate, both of PA-ACS grade, were supplied by Panreac (Barcelona, Spain). 2,7-diclorofluorescein of analytical grade was supplied by Fluka Chemical Co. (Ronkonkoma, NY, USA). 5α-cholestane-3β-ol and betulin were supplied by Fluka Chemical Co. (Ronkonkoma, NY, USA) and used as internal standards. Derivatizing reagent, a mixture 99:1 (v/v) of N,O-bis (trimethylsilyl)-trifluoroacetamide and trimethylchlorosilane were supplied by Tokyo Chemical Industry CO. (Tokyo, Japan). Anhydrous pyridine of analytical grade (ref. 7463) was supplied by Merck (Darmstadt, Germany). Diatomaceous Earth, 6/60 mesh was supplied by Restek (Bellefonte, PA, USA). Adsorbent-phase Bondesil-NH2 40 μm was supplied by Varian, Inc., (Walnut Creek, CA, USA). Commercial SLE cartridge Strata DE, 60 cc, was supplied by Phenomenex (Torrance, CA, USA). All other reagents were of analytical grade.

2.2. Samples

Virgin olive oil from the cultivar variety Picual and refined olive pomace oil were used. The virgin olive oil sample was obtained from the oil mill pilot plant located in the "Instituto de la Grasa (CSIC)", operating in the usual conditions, during the season 2018/2019. Refined olive pomace oil and refined sunflower seed oil samples were supplied by a local refining industry. The olive oil samples were homogenized and divided into aliquots to carry out the different assays. Additionally, a refined sunflower oil sample was purchased from a local store. The unsaponifiable matter was extracted using the different methods described in the sections below.

2.3. Instrumentation

2.3.1. HPLC Isolation

Sterols and dialcoholic triterpenic fractions were isolated by HPLC. The HPLC system consisted of an Agilent (Palo Alto, CA, USA) 1200 series liquid chromatograph, equipped with a micro vacuum degasser, a binary pump, an auto-sampler injector provided with a preparative-head assembly of 900 μL, a Peltier furnace, a refractive index detector 1100 series and an analytical fraction collector installed at the exit of the detector for the recovery of the sterol fraction. A chemical station HP was used for controlling and monitoring the system. The separation was performed in a 150 mm × 3.9 mm, particle size 4 μm Nova Pak Silica 60 Å Water Millipore Corporation (Milford, MA, USA) column. The temperature of the column and the detector were held, respectively, at 20 and 35 °C. The mobile phase was n-hexane/ethyl acetate 90/10 (v/v). The flow rate was established at 0.6 mL·min^{-1} for 30.00 min.

2.3.2. Gas Chromatography-Flame Ionization Detector (GC-FID) Analysis

The collected fraction using the HPLC system was analyzed in an Agilent (Palo Alto, CA, USA) 7890A gas chromatograph equipped with a split/splitless injector and a flame ionization detector; a capillary HP-5MS column (30 m × 0.25 mm I.D., 0.25 μm film thickness, Agilent J & W, Palo Alto, CA, USA) and an Agilent G 4513A automatic injector were used. The oven temperature was kept at 265 °C isothermally. The operating condition of injector was split mode and its temperature was kept at 310 °C, while the detector temperature was 310 °C and the injection volume was 1 µL. Hydrogen was used as the carrier gas at 1.0 mL·min^{-1} in constant flow mode and a split ratio of 1:10. Air and hydrogen at flow rates of 300 and 30 mL·min^{-1}, respectively, were used for the detector, which had an auxiliary flow of 30 mL·min^{-1} of nitrogen.

2.3.3. ATR-FTIR Spectroscopy

A Bruker 55 Equinox S FTIR spectrometer with a DGTS detector (Bruker Optics, Ettlingen, Germany) was used in this study. The sampling station was equipped with an overhead, detachable attenuated total reflectance (ATR, six bounces, Specac, Orpington, UK) accessory consisting of a zinc selenide crystal mounted in a shallow channel for the sample containment. Each spectrum was recorded at room temperature in the region of 4000–600 cm^{-1} by an average of 50 scans at a resolution of 4 cm^{-1}.

2.4. Sample Treatment

2.4.1. IOC and EU Methods (Liquid–Liquid Extraction Method)

The preparation and analysis of the unsaponifiable matter were carried out in accordance with the IOC and EU methods of analysis (official methods) [5–7]. The oil samples, with added 5α-cholestane-3β-ol as an internal standard, were saponified with potassium hydroxide 2 M in ethanolic solution (with 20% of water), and the unsaponifiable fractions were then extracted three times with diethyl ether.

The 4-desmethylsterol and triterpenic dialcohol fractions were extracted, as has been previously described in the literature [5–7]. Briefly, 5.00 ± 0.10 g of the oil sample was weighed in a flask containing 5α-cholestane-3β-ol (0.5 mL of a solution of 0.01% m/v in ethyl acetate was previously added and evaporated until dryness). Then, the oil sample containing the internal standard was saponified for 60 min with 50 mL of 2 M ethanolic potassium hydroxide with 20% water. The solution was passed into a 500 mL decanting funnel, 100 mL of distilled water was added and the mixture was extracted twice with three 80 mL portions of diethyl ether. The organic extracts were combined in another funnel and washed several times with 100 mL portions of water until the wash reached neutral pH. The diethyl ether solution was dried over anhydrous sodium sulphate and evaporated to dryness in a rotary evaporator at 30 °C under reduced pressure.

The complete unsaponifiable dried fraction was then redissolved in approximately 3.00 mL of the mobile phase, and 250 µL of the solution was injected into the HPLC system as described in the Instrumentation section. Subsequently, the fraction that eluted from minutes 11.00 to 24.00 was recovered through the analytical fraction collector. The solvent was evaporated to dryness under reduced pressure. The 4-desmethylsterol and triterpenic dialcohol fractions were treated with 150 µL of the derivatizing reagent to obtain the trimethyl silyl derivates for subsequent GC-FID analysis.

2.4.2. Proposed Method (Supported Liquid Extraction Method)

The internal standard solution (40 µL of α-cholestanol for virgin olive oil and 100 µL for refined olive pomace oil, and 100 µL of Betulin for both samples) was introduced into a 4 mL vial and the solvent was evaporated under a N2 stream. Next, 0.300 ± 0.010 g of the virgin oil sample or 0.200 ± 0.010 g of the olive pomace oil were weighed in the same vial which contained the standard. One milliliter of an ethanolic solution of KOH 2 M was added to the vial, and it was closed and heated up in a thermo-block for 45 min at 85 °C, shaking the vial every 15 min.

Once the time was over, 2 mL of distillate water was added to the vial and the content was poured into a homemade prepared column of 20 mm I.D. filled with 1 g of amine (lower layer) and 5 g of diatomaceous earth (upper layer) of particle size <0.5 mm, obtained by sifting as described in the Reagents and Solutions section, leaving the mixture for at least ten minutes before eluting. Then, a volume of 45 mL of a hexane:ethyl acetate mixture (85:15, v/v) was passed through the cartridge. The whole content was collected in a 50 mL flask and evaporated to dryness in a rotary evaporator at 30 °C under reduced pressure; then, it was dissolved again with 300 µL of the mobile phase. Next, the content was centrifuged at 14,000× g for approximately 1 min. The upper layer was collected with a pipette, poured into a HPLC vial and 250 µL was injected into the chromatograph. Subsequently, as described in the IOC or EU method, the fraction that eluted from minutes 11.00 to 24.00 was collected, evaporated, silylated with 150 µL of derivatizing reagent, and injected into the GC-FID.

Figure 1 details the whole procedure for the determination of sterol and triterpenic dialcohol fractions using the proposed method: first of all, the saponification was performed, then the sample extraction by SLE, HPLC isolation, derivatization and, finally, GC-analysis.

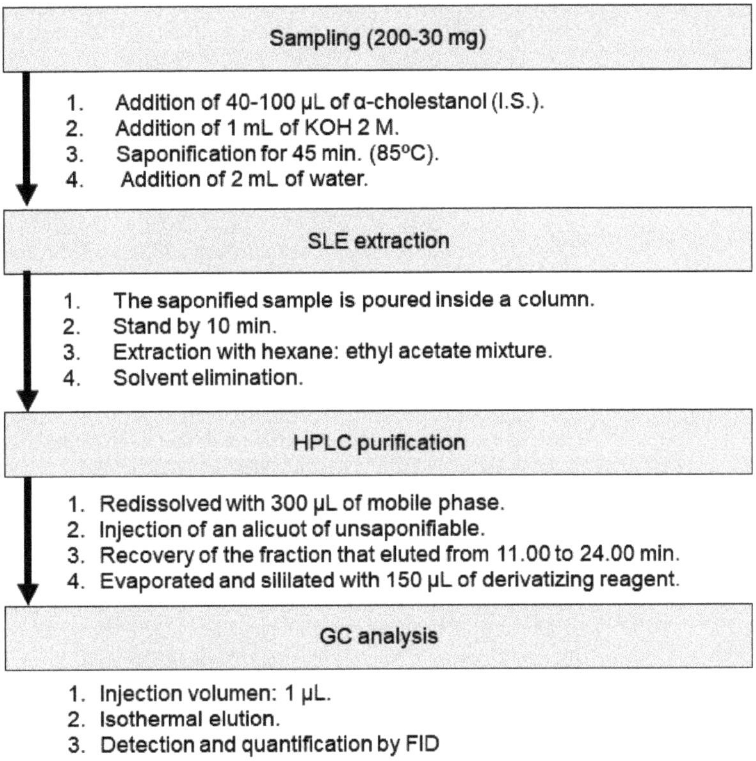

Sampling (200-30 mg)
1. Addition of 40-100 µL of α-cholestanol (I.S.).
2. Addition of 1 mL of KOH 2 M.
3. Saponification for 45 min. (85 °C).
4. Addition of 2 mL of water.

SLE extraction
1. The saponified sample is poured inside a column.
2. Stand by 10 min.
3. Extraction with hexane: ethyl acetate mixture.
4. Solvent elimination.

HPLC purification
1. Redissolved with 300 µL of mobile phase.
2. Injection of an alicuot of unsaponifiable.
3. Recovery of the fraction that eluted from 11.00 to 24.00 min.
4. Evaporated and sililated with 150 µL of derivatizing reagent.

GC analysis
1. Injection volumen: 1 µL.
2. Isothermal elution.
3. Detection and quantification by FID

Figure 1. Workflow for the determination of sterols and triterpenic dialcohols in olive or olive pomace oils.

2.5. GC-Data Analysis

The GC-peak areas were calculated with Agilent ChemStation OpenLAB, and the determination of individual 4-desmethylsterols and triterpenic dialcohols was carried out by evaluating the corresponding relative percentage according to the normalization area procedure assuming an equal factor response for any species. The quantitative determination of the total sterols and triterpenic dialcohols were performed relative to the peak area of the known concentration of the internal standard.

2.6. ATR-FTIR Spectra

IR spectra were acquired for the unsaponifiable matter obtained by the official method as well as by the suggested method, according to the described method by Tena et al. [26].

2.7. Recovery and Precision

In order to study the recovery and reproducibility of the present method, a complementary experiment was carried out. Recovery data were calculated, comparing the results obtained from the proposed and the official methods. Six replicates were made in each case. For the determination of reproducibility, the replicates were performed on different days and in the same laboratory [27].

3. Results and Discussion

Among all the possibilities for the determination of sterols in vegetable oils, gas chromatography with a FID detector is the last step in the process of their quantification and, it is, with minor differences, common to all proposals. However, the previous steps are key to performing their determination accurately, and there are important differences between the different proposed methods as well as with respect to the official methods.

3.1. Saponification, SLE and HPLC Isolation

The first step for the sterols' determination, the saponification, is mandatory in all of the methodologies described, as these unsaponifiable compounds are present in a free, and esterified with fatty acids, form. Therefore, for their whole determination, they must be transformed and isolated in the form of free sterols.

In the regulations, after one hour of boiling, the unsaponifiable matter can be isolated from the saponified one. For that purpose, the official methods [5–7] described several liquid–liquid extractions with diethyl ether where the unsaponifiable fraction is extracted almost free of the saponified part, which is solubilized into the water. Next, a cleaning step of the collected diethyl ether fractions with distilled water is necessary to guarantee the absence of the basic reagent used for the saponification process and the almost complete absence of the saponified matter. As can be deduced, large volumes of solvent and time are consumed. The proposed method here described starts with the saponification of an oil sample about sixteen times smaller than that used in the official method; it is saponified for 45 min and is passed through a SLE column.

Table 1 shows the details of the sample quantities, water addition, volumes, washes, drying of the sample and purification that are carried out in the proposed procedure compared to the regulation and to another process recommended by a commercial company [17]. Among others, the main differences are the sample amounts and the solvent used for the extractions. As can be seen, when using SLE, the volume of the sample is reduced by 12–25 times compared to L–L, and the volume of the solvent is also reduced by more than six times. All this makes the official method long and tedious, where, in addition, the formation of emulsions that must be broken is frequent.

Table 1. Details of the procedures carried out for the isolation of the sterol and triterpenic dialcohol fractions.

Steeps	L–L Method	SLE Methods	
	Official Methods	Commercial [17]	Proposed Method
Sample amount	5 g	0.2–0.4 g	0.3 g
Water addition	100 mL	13.5 mL	2 mL
Extraction	3 × 100 mL	3 × 15 mL	45 mL
Washed	3 × 50 mL	no	no
Dried	30–50 g Na_2SO_4	SPE Na_2SO_4	no
FFAs removal	KOH in TLC Si	KOH in SPE Si	no
Purification	TLC or HPLC	no	HPLC

For all of this, the SLE is, at present, a rapid and the best alternative to the L–L extraction used in the official method for isolating the unsaponifiable matter. Thus, Table 2 presents a detailed comparison of the operation times used in each of the steps in the L–L (official methods) and SLE methods; also, the proposed method is compared to that proposed by the commercial company [17]. As can be observed, and considering all the steps, the new procedure is more than twice as fast as the official one, which is an important advance in work per day, and 1.66 times faster than the current commercial method [17]. Furthermore, as Table 1 shows, the number of steps in the suggested method is reduced to 57%, regarding the official method, which notably reduces losses and pollution during the process.

Table 2. Comparison of the operation times (min.) for the L–L and SLE methods.

Steeps	L–L Method	SLE Methods	
	Official Methods	Commercial [17]	Proposed Method
Sampling and Internal standard addition	7:00	1:20	1:20
KOH 2M training and addition	3:11	1:00	1:00
Saponification	60:00	50:00	45:00
Water addition and cooling	30:00	15:00	10:00
Extraction with organic solvent	15:00	5:00	5:00
Washed	20:36	no	no
Dryed	20:36	15:00 SPE Na_2SO_4	no
Free Fatty Acids removal	17:25	15:00 SPE Si, KOH	no
Purified + derivatization	50:00	30:00	30:00
GC Analysis	30:00	70:00	30:00
TOTAL time	253:48	202:20	122:20

The SLE used here combines diatomaceous earth with particle size <0.5 mm with a layer of amino phase in the same cartridge. This blend guarantees an unsaponifiable matter sample free of water and free fatty acids, which is of great importance for the good isolation of each unsaponifiable component and particularly the sterols and triterpenic dialcohols in the following step. On the other hand, the use of diethyl ether, acting as an extraction solvent as proposed in the regulation and commercial methods, extracts the unsaponifiable matter together with some soaps, which interfere in the purification of the unsaponifiable matter and must be eliminated with water addition. Nevertheless, this inconvenience is avoided by using the admixture hexane:ethyl acetate here proposed.

Once the unsaponifiable matter has been obtained, it is necessary to isolate, by means of a chromatographic technique (TLC or HPLC), the series of compounds of interest—in our case, the sterols and triterpenic dialcohols. This step is almost mandatory to obtain better precision in the quantification, as other unsaponifiable compounds, such as alcohols, tocopherols or hydrocarbons, interfere.

In the official regulations, TLC has been used for years as the best method for the isolation, but at present, an HPLC method is also proposed for it. Before its recent inclusion in the regulations of the European Union and IOC in 2020 [5,7], the HPLC method for sterol isolation was studied by various different authors [1,16,28].

With the selected conditions studied here, in the HPLC chromatogram of the unsaponifiable matter corresponding to the oil samples, the fraction collected ranged from Δ^5- and Δ^7-sterols to erithrodiol+Uvaol.

The RP-HPLC method proposed for the isolation is more precise and less time-consuming, and many other factors indicate that the HPLC procedure is better than TLC. Thus [29], reported that the insufficient separation in TLC between the band of sterols and triterpenic alcohols, the delimitation of the band of sterols in the TLC and the impurities close to the band of triterpenic dialcohols are the main causes for the lack of precision in the determination of Δ^7-sterols.

On the other hand, the eluent admixture hexane-diethyl ether (65:35 v/v) at the flow rate proposed by official methods (EU, IOC) [16] was replaced here by n-hexane/ethyl acetate 90/10 (v/v) at a flow rate of 0.6 mL·min^{-1}; the use of diethyl ether presents high pressure and burble problems in the HPLC equipment.

The amount and size of diatomaceous earth particles used to prepare the columns is extremely important, due to the fact that these parameters determine the amount of absorbed water and water flux through the column; in the specific case of the proposed method, 5 g of diatomaceous earth with a particle size <0.05 mm was used. Table 3 shows the differences between the SLE columns in the suggested and in the commercial method [17]; the amount of diatomaceous earth used in the suggested method is only 5 g, compared to 19 g of this material in the commercial columns, which is three times greater in volume.

Table 3. Comparison between SLE columns.

	Commercial [17]	Proposed Method
Volume (mL)	75	25
Stuffing amount (g)	19.0	5.0
Length (cm)	14.0	8.5
Sorbent	Unknown	diatomaceous earth

This newly suggested method does not require the drying and free fatty acid removal steps, in contrast to methods described in the literature [17,25], because using ethyl acetate and hexane in conjunction with the type of column minimizes the amount of water and soap extracted to trace levels. Regarding the removal of possible free fatty acids from hydrolysis of extracted soap traces, it takes place in the same SLE column as that in which 1 g of adsorbent phase Bondesil-NH$_2$ 40 µm is deposited. In order to check what is mentioned above, ATR-FTIR spectra were examined for unsaponifiable compounds of a virgin olive oil obtained via the official method as well as via the suggested method, in both cases before the purification step by HPLC. As Figure 2 shows, for the unsaponifiable matter obtained through the official method, there is a high-intensity band around a wavelength of 1716 cm^{-1} in the infrared spectra, which corresponds to the free fatty acids [26,30], whereas this band is almost negligible for the unsaponifiable matter obtained through the suggested method.

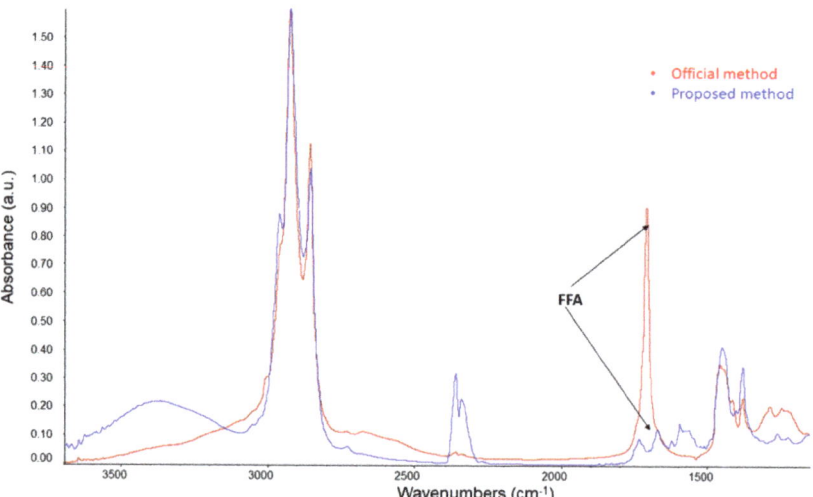

Figure 2. Spectra ATR—FTIR of the unsaponifiable matter from virgin olive oil obtained by official method and proposed method.

The HPLC technique was used to purify the unsaponifiable fraction. A chromatographic or solid phase extraction technique [5,16,25] is needed in the cases of olive oil and olive pomace oil in order to remove interferences that might appear when analyzing, through the GC technique, the sterol fraction with specific triterpenic alcohols and methyl sterols, among other reasons. However, the official methods do not use any purification techniques before HPLC isolation.

The official methods use mixtures of hexane/diethyl ether 50:50 (v/v) as the mobile phase; using diethyl ether in HPLC may cause bubble and pressure oscillation issues. However, these issues may be avoided by using mixtures of hexane/ethyl acetate 90:10 (v/v) at lesser fluxes, as in the suggested method. Moreover, Si columns with a particle size of 5 µm and dimensions of 250 mm × 4.6 mm might be substituted by columns with dimensions of 150 mm × 3.9 mm and a particle size of 4 µm; thanks to this modification, resolution is improved, obtaining a lesser peak width.

Figure 3 shows a HPLC chromatogram of the unsaponifiable fraction from the refined olive pomace oil sample according to the suggested method. As can be observed, four groups of compounds are clearly differentiated: aliphatic and terpenic hydrocarbons (1), linear and triterpenic alcohols and methyl sterols (2), sterols (3) and triterpenic dialcohols (4). Each one can be recovered in an established time interval; in this work, only the sterol and triterpenic dialcohol fractions were studied, which may be recovered between 11 and 24 min without interferences from methyl sterols and triterpenic alcohols. This interval is relatively similar to that suggested in the official methods; however, as can be observed in the figure, the resolution between the different groups of compounds is better in the case of the suggested method, because the unsaponifiable matter is not purified in the official methods to guarantee the absence of free fatty acids.

Figure 4 shows a HPLC chromatogram of the unsaponifiable fraction from refined sunflower seed oil. As this figure shows, within the chromatographic conditions proposed in the present work, a good resolution between sterols with Δ^5- and Δ^7-sterols structure can be obtained.

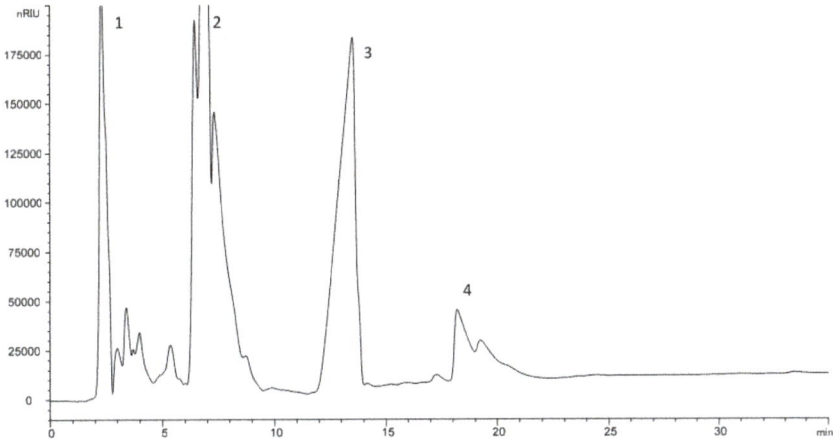

Figure 3. HPLC chromatogram of the unsaponifiable fraction from olive pomace oil. 1: Aliphatic and terpenic hydrocarbons; 2: linear and triterpenic alcohols and methyl sterols; 3: sterols; 4: triterpenic dialcohols.

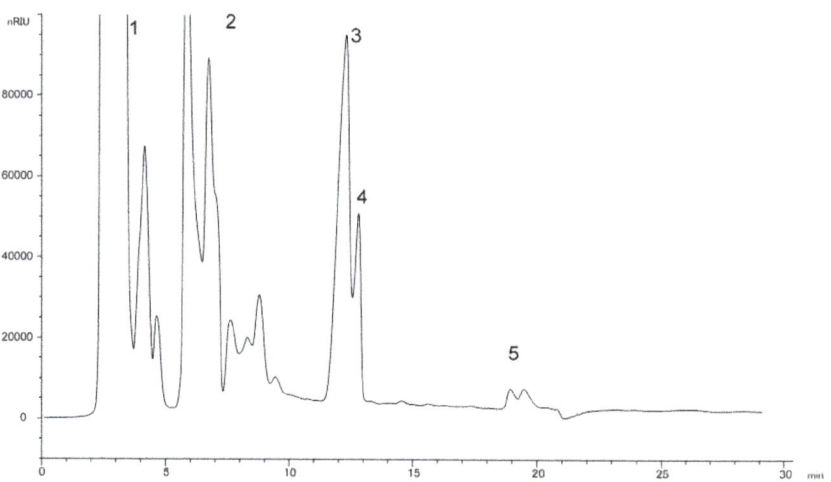

Figure 4. HPLC chromatogram of unsaponifiable fraction from a mixture of olive pomace oil and sunflower seed oil. 1: Aliphatic and terpenic hydrocarbons; 2: linear and triterpenic alcohols and methyl sterols; 3: Δ^5-sterols; 4: Δ^7-sterols; 5: triterpenic dialcohols.

3.2. Gas Chromatography Determination of Sterols and Dialcoholic Triterpenes

Figure 5 shows a gas-chromatogram corresponding to sterols and triterpenic dialcohols obtained from a refined olive pomace oil. Sixteen peaks are numbered and correspond to the compounds, which eluted in the same order as specified in Table 4. The chromatographic profile matches with the chromatograms published in the regulations for virgin and refined olive oils, and as the conditions are the specified and recommended in the regulation, β-sistosterol elutes in the range of 20 ± 5 min. The internal standard allows the quantitative determination of all the sterols and triterpenic dialcohols, and is well separated from all the peaks, including the nearest, the cholesterol eluting just ahead of it. It is noteworthy that it is a refined olive oil, since just before the clerosterol peak, the $\Delta^{5,23}$-stigmastadienol elutes, only present when the oil has been refined. In addition, the TD, erythrodiol and uvaol are undoubtedly identified by their magnitude, as corresponds to an olive pomace oil, but also because their relative retention times with respect to β-sistosterol are 1.41 and 1.52 for erythrodiol and uvaol, respectively, as specified in the regulation.

If any of the above-mentioned purification techniques are not employed, it is hard to avoid interferences. Some authors have suggested a longer analysis time in GC in order to avoid such interferences [17]. However, this procedure has low efficiency because it increases the retention of the compounds inside the column, increasing diffusion and, therefore, reducing sensibility and precision.

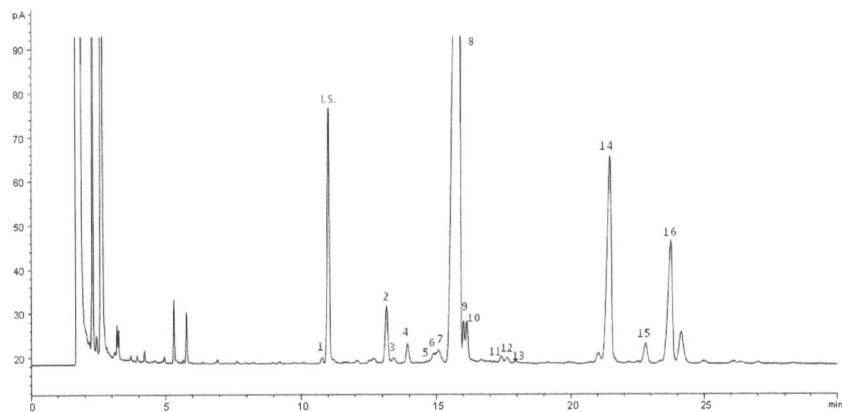

Figure 5. GC—FID chromatogram of olive pomace oil sterols fraction recollected from HPLC chromatograph. 1: cholesterol; 2: campesterol; 3: campestanol; 4: stigmasterol; 5: Δ^7-campesterol; 6: $\Delta^{5,23}$-stigmastadienol; 7: clerosterol; 8: β-sitosterol; 9: sitostanol; 10: Δ^5-avenasterol; 11: $\Delta^{5,24}$-stigmastadienol; 12: Δ^7-stigmastenol; 13: Δ^7-avenastol; 14: erythrodiol; 15: uvaol; 16: betulin.

Table 4. Reproducibility of the sterols and triterpenic dialcohols of oil samples by the proposed SLE method and official methods.

	VOO						OPO					
	SLE			OM *			SLE			OM *		
	Means ** N = 6	SD	RSD (%)	Means ** N = 6	SD	RSD (%)	Means ** N = 6	SD	RSD (%)	Means ** N = 6	SD	RSD (%)
Cholesterol	nd	nd	nd	nd	nd	nd	0.20	0.03	15.00	0.15	0.03	20.36
Campesterol	2.83	0.07	2.47	2.81	0.04	1.42	2.63	0.03	1.07	2.56	0.01	0.50
Stigmasterol	0.74	0.07	9.46	0.68	0.05	7.35	0.89	0.06	7.01	0.95	0.02	1.61
$\Delta^{5,23}$-stigmastadienol	nd	nd	nd	nd	nd	nd	0.55	0.02	3.93	0.51	0.00	0.88
Clerosterol	1.04	0.05	4.81	1.06	0.1	9.43	0.86	0.02	1.89	0.86	0.02	2.42
β-sitosterol	83.95	1.07	1.27	83.96	1.75	2.08	74.75	0.20	0.27	73.85	0.45	0.60
Sitostanol	0.69	0.03	4.35	0.78	0.04	5.13	1.45	0.11	7.38	1.44	0.06	4.23
Δ^5-avenasterol	7.37	0.18	2.44	7.34	0.18	2.45	1.48	0.03	2.27	1.49	0.12	7.87
$\Delta^{5,24}$-stigmastadienol	0.70	0.04	5.71	0.61	0.07	11.48	1.31	0.04	2.81	1.46	0.02	1.71
Δ^7-stigmastenol	0.42	0.04	9.52	0.38	0.04	10.53	0.35	0.03	8.57	0.31	0.04	12.90
Δ^7-avenasterol	0.74	0.07	9.46	0.66	0.07	10.61	0.28	0.01	3.57	0.35	0.01	2.13
Erythrodiol + Uvaol ***	1.52	0.06	3.95	1.72	0.04	2.33	15.25	0.29	1.88	16.07	0.56	3.51
Total sterols (ppm)	1231.38	30.21	2.45	1224.19	30.73	2.51	5529.87	309.19	5.59	5183.04	57.45	1.11

VOO: Virgin olive oil; OPO: olive pomace oil; * official methods; nd: non detected; ** the mean data are presented as percentages of total sterols, without include the triterpenic dialcohols; *** the mean data are presented as percentages of total sterols; SD: Standard deviation; RSD: Relative standard deviation.

3.3. Validation of the Method

The proposed procedure was validated in house by performing six replicated analyses of two different olive oil samples. The mean and the standard deviation were calculated, and the data are reported in Table 4. Furthermore, accuracy was evaluated, and the results are also included. It can be seen that the repeatability is good in all cases because the values of the relative standard deviations are lower than the reference value derived from the Horwitz equation [31] (RSDH = 16.4%). Therefore, the results for different sterols and triterpenic dialcohols indicate a good repeatability for the assay.

In order to evaluate the recovery of the proposed method, the mean values of the different compounds obtained from six replicates were compared to the results obtained by the procedure specified in the official methods (Table 5). As can be observed in the mentioned table, all compounds showed recovery values higher than 88.00%, which means a good recovery was obtained in all cases.

Table 5. Recovery percentages of the sterols and triterpenic dialcohols of VOO and OPO, in relation to the official method (100%). ($N = 6$).

	VOO	OPO
Cholesterol	nd	133.33
Campesterol	100.71	102.73
Stigmasterol	108.82	93.68
$\Delta^{5,23}$-stigmastadienol	nd	107.84
Clerosterol	98.11	100.00
β-sitosterol	99.99	101.22
Sitostanol	88.46	100.69
Δ^5-avenasterol	100.41	99.33
$\Delta^{5,24}$-stigmastadienol	114.75	89.73
Δ^7-stigmastenol	110.53	112.90
Δ^7-avenasterol	112.12	80.00
Erytrodiol + Uvaol	88.37	94.90
Total sterols (ppm)	100.59	106.69

nd: non detected.

4. Conclusions

The analytical method proposed for isolating and quantifying the unsaponifiable fraction is based on SLE extraction, and it may be proposed as a good alternative to the liquid–liquid extraction methods (official methods), since it offers a clear advantage over the recovery of the sterol and triterpenic dialcohol fractions, and the results obtained are not significantly different from those obtained by the official methods.

The new SLE/HPLC method detailed in this work allows a rapid and highly accurate separation of the different compound families that are part of the unsaponifiable matter from olive oils. The isolation of the unsaponifiable fraction proposed makes the analysis of the compounds included in this fraction (sterols, aliphatic alcohols, tocopherols, etc.) easier and less time-consuming than those previously reported. Thus, the analysis time was reduced by more than half, and the volume of solvent used was also reduced by more than six times with respect to the official methods.

It has been demonstrated that the methodology based on an off-line combination of HPLC and GC-FID is a good, quick and reproducible analytical method for the isolation and quantification of the sterols and triterpenic alcohols in olive oils. The precision and accuracy of the procedure described have been checked, showing a high recovery of the different compounds studied.

Furthermore, this method ensures the removal of fatty acids, avoiding all the possible interferences during the GC quantification.

The results obtained enable the assessment of the olive oil's quality in accordance with UE and IOC sterol criteria and agree with the mentioned regulations.

Author Contributions: Conceptualization, M.L.-C. and M.d.C.P.-C.; methodology, M.L.-C. and M.d.C.P.-C.; formal analysis, M.L.-C. and M.d.C.P.-C.; investigation, M.L.-C. and M.d.C.P.-C.; resources, M.L.-C. and M.d.C.P.-C.; writing—original draft preparation, M.L.-C. and M.d.C.P.-C.; writing—review and editing, M.L.-C. and M.d.C.P.-C. All authors have read and agreed to the published version of the manuscript.

Funding: This research received no external funding.

Data Availability Statement: Not applicable.

Acknowledgments: The authors are grateful to E. Rubio López for the technical assistance and D. L. García-González for performing the ATR-FTIR analysis.

Conflicts of Interest: The authors declare no conflict of interest.

References

1. León-Camacho, M.; Morales, M.T. Chapter 7: Gas and Liquid Chromatography: Methodology Applied to Olive Oil. In *Handbook of Olive Oil, Analysis and Properties*, 1st ed.; Aparicio, R., Harwood, J., Eds.; Aspen Publishers: Gaithersburg, MD, USA, 2001; pp. 163–217.
2. Azadmard-Damirchi, S. Review of the use of phytosterols as a detection tool for adulteration of olive oil with hazelnut oil. *Food Addit. Contam.* **2010**, *27*, 1–10. [CrossRef]
3. Kamm, W.; Dionisi, F.; Hischenhuber, C.; Engel, K. Authenticity assessment of fats and oils. *Food Rev. Int.* **2001**, *17*, 249–290. [CrossRef]
4. Dieffenbacher, A.; Pocklington, W.D. (Eds.) IUPAC. 2.401: Determination of the Unsaponifiable Matter in Standard Methods for the Analysis of Oil, Fats and Derivatives, 7th ed; Blackwell Scientific Publications: Oxford, UK, 1987.
5. International Olive Council. *Determination of the Sterol Composition and Content and Alcoholic Compounds by Capillary Gas Chromatography*; COI/T.20/ Doc. No 26/Rev.; International Olive Council: Madrid, Spain, 5 June 2020.
6. Regulation (EEC) No 2568. Of 11 July 2011. On the Characteristics of Olive Oil and Olive-Residue Oil and on the Relevant Methods of Analysis (Consolidated Text). *Off. J. Eur.* **1991**, *L 248*, 35–46.
7. Regulation (EU) No 1604. Of 27 September 2019. Amending Regulation (EEC) No 2568/91 on the characteristics of olive oil and olive-residue oil and on the relevant methods of analysis. *Off. J. Eur.* **2019**, *L 250*, 36–48.
8. International Olive Council. *Trade Standard Applying to Olive Oils and Olive Pomace Oils*; COI/T.15/NC No 3/Rev.; International Olive Council: Madrid, Spain, 16 June 2021.
9. Dieffenbacher, A.; Pocklington, W.D. (Eds.) IUPAC. 2.403: Identification and Determination of Sterols by Gas Liquid Chromatography in Standard Methods for the Analysis of Oil, Fats and Derivatives, 7th ed.; Blackwell Scientific Publications: Oxford, UK, 1987.
10. Dieffenbacher, A.; Pocklington, W.D. (Eds.) IUPAC. 2.404: Determination of the Total Sterols Content in Standard Methods for the Analysis of Oil, Fats and Derivatives, 7th ed.; Blackwell Scientific Publications: Oxford, UK, 1987.
11. Toivo, J.; Piiroren, V.; Kalo, P.; Varo, P. Gas chromatographic determination of major sterols in edible oils and fats using solid-phase extraction in sample preparation. *Chromatographia* **1998**, *48*, 745–750. [CrossRef]
12. Azadmard-Damirchi, S.; Dutta, P.C. Novel solid-phase extraction method to separate 4-desmethyl-, 4-monomethyl-, and 4,4′-dimethylsterols in vegetable oils. *J. Chrom. A* **2006**, *1108*, 183–187. [CrossRef] [PubMed]
13. Bello, A.C. Rapid isolation of the sterol fraction in edible oils using silica cartridge. *J. AOAC Int.* **1992**, *75*, 1120–1123. [CrossRef]
14. Señoráns, F.J.; Tabera, I.; Herraiz, M. Rapid separation of free sterols in edible oils by on-line coupled reversed phase liquid chromatography-gas chromatography. *J. Agric. Food Chem.* **1996**, *44*, 3189–3192. [CrossRef]
15. Brewington, C.R.; Caress, E.A.; Schwartz, D.P. Isolation and identification of new milk fat. *J. Lipid Res.* **1979**, *11*, 355–361. [CrossRef]
16. Amelio, M.; Rizzo, R.; Varazini, F. Determination of sterols, erythrodiol, uvaol and alkanols in olive oils using combined solid-phase extraction, high-performance liquid chromatographic and high-resolution gas chromatographic techniques. *J. Chrom. A.* **1992**, *606*, 179–185. [CrossRef]
17. Chapa, E.; Brusius, M.; Krepich, S.; Aqeel, Z.; Detsch, J. Determination of Sterols in Olive Oil Using Supported Liquid Extraction (SLE), Solid Phase Extraction (SPE) and GC-FID, (TN-1004). 2017. Available online: https://www.phenomenex.com/ViewDocument?id=determination+of+sterols+in+olive+oil&fsr=1 (accessed on 29 June 2018).
18. Grob, K. Development of the transfer techniques for on-line high-performance liquid chromatography-capillary gas chromatography. *J. Chrom. A* **1995**, *703*, 265–276. [CrossRef]
19. Vreuls, J.J.; de Jong, G.J.; Ghijsen, R.T.; Th Brinkman, U.A. Liquid chromatography coupled on-line with gas chromatography: State of the art. *J. AOAC. Int.* **1994**, *77*, 306–327. [CrossRef]
20. Señoráns, F.J.; Herraiz, M.; Tabera, J. On-line reversed-phase liquid chromatography-capillary gas chromatography using a programmed temperature vaporizer as interface. *J. Sep. Sci.* **1995**, *18*, 433–438. [CrossRef]
21. Señoráns, F.J.; Reglero, G.; Herraiz, M. Use of a Programmed Temperature Injector for On-Line Reversed-Phase Liquid Chromatography-Capillary Gas Chromatography. *J. Chromatogr. Sci.* **1995**, *33*, 446–450. [CrossRef]
22. Homberg, E. Vitamin D-Bestimmung in Lebertram. *Fat Sci. Tech.* **1993**, *95*, 228–230.
23. Hadorn, H.; Zürcher, K. Der Scheidetrichter, ein mangelhaftes Gerät im analytischen Labor. *Gordian* **1973**, *73*, 198–201.
24. Johnson, C.R.; Zhang, B.; Fantauzzi, P.; Hocker, M.; Yager, K.M. Libraries of N-alkylaminoheterocycles from nucleophilic aromatic substitution with purification by solid supported liquid extraction. *Tetrahedron* **1998**, *54*, 4097–4106. [CrossRef]
25. Mathison, B.; Holstege, D. A Rapid Method to Determine Sterol, Erthyrodiol, and Uvaol Concentrations in Olive Oil. *J. Agric. Food Chem.* **2013**, *61*, 4506–4513. [CrossRef] [PubMed]
26. Tena, N.; Aparicio, R.; García-González, D.L. Thermal Deterioration of Virgin Olive Oil Monitored by ATR-FTIR Analysis of Trans Content. *J. Agric. Food Chem.* **2009**, *57*, 9997–10003. [CrossRef]
27. Thompson, M.; Ellison, S.L.R.; Wood, R. Harmonized guidelines for singlelaboratory validation of methods of analysis. *Pure Appl. Chem.* **2002**, *74*, 835–855. [CrossRef]
28. Cert, A.; Moreda, W.; García-Morena, J. Determinación de esteroles y dialcoholes triterpénicos en aceite de oliva mediante separación de la fracción por cromatografía líquida de alta eficacia y análisis por cromatografía de gases. Estandarización del método analítico. *Grasas Aceites* **1997**, *48*, 207–218. [CrossRef]

29. León-Camacho, M.; Cert Ventulá, A. Recomendaciones para la aplicación de algunos métodos analíticos incluidos en el reglamento CEE 2568/91 relativo a las características de los aceites de oliva y de orujo de oliva. *Grasas Aceites* **1994**, *45*, 395–401. [CrossRef]
30. Tena, N.; Aparicio-Ruiz, R.; García-González, D.L. Time Course Analysis of Fractionated Thermoxidized Virgin Olive Oil by FTIR Spectroscopy. *J. Agric. Food Chem.* **2013**, *61*, 3212–3218. [CrossRef] [PubMed]
31. European Commission Decision No 657. Of August 12, 2002, implementing Council Directive 96/23/EC concerning the performance of analytical methods and the interpretation of results. *Off. J. Eur.* **2002**, *L 221*, 8–36.

Article

Authenticity Assessment and Fraud Quantitation of Coffee Adulterated with Chicory, Barley, and Flours by Untargeted HPLC-UV-FLD Fingerprinting and Chemometrics

Nerea Núñez [1,*], Javier Saurina [1,2] and Oscar Núñez [2,3,*]

[1] Department of Chemical Engineering and Analytical Chemistry, University of Barcelona, Martí i Franquès 1-11, E08028 Barcelona, Spain; xavi.saurina@ub.edu

[2] Research Institute in Food Nutrition and Food Safety, University of Barcelona, Recinte Torribera, Av. Prat de la Riba 171, Edifici de Recerca (Gaudí), Santa Coloma de Gramenet, E08921 Barcelona, Spain

[3] Serra Húnter Fellow, Generalitat de Catalunya, E08007 Barcelona, Spain

* Correspondence: nereant7@gmail.com (N.N.); oscar.nunez@ub.edu (O.N.)

Citation: Núñez, N.; Saurina, J.; Núñez, O. Authenticity Assessment and Fraud Quantitation of Coffee Adulterated with Chicory, Barley, and Flours by Untargeted HPLC-UV-FLD Fingerprinting and Chemometrics. *Foods* **2021**, *10*, 840. https://doi.org/10.3390/foods10040840

Academic Editor: Raúl González-Domínguez

Received: 25 March 2021
Accepted: 9 April 2021
Published: 12 April 2021

Publisher's Note: MDPI stays neutral with regard to jurisdictional claims in published maps and institutional affiliations.

Copyright: © 2021 by the authors. Licensee MDPI, Basel, Switzerland. This article is an open access article distributed under the terms and conditions of the Creative Commons Attribution (CC BY) license (https://creativecommons.org/licenses/by/4.0/).

Abstract: Coffee, one of the most popular drinks around the world, is also one of the beverages most susceptible of being adulterated. Untargeted high-performance liquid chromatography with ultraviolet and fluorescence detection (HPLC-UV-FLD) fingerprinting strategies in combination with chemometrics were employed for the authenticity assessment and fraud quantitation of adulterated coffees involving three different and common adulterants: chicory, barley, and flours. The methodologies were applied after a solid–liquid extraction procedure with a methanol:water 50:50 (v/v) solution as extracting solvent. Chromatographic fingerprints were obtained using a Kinetex® C18 reversed-phase column under gradient elution conditions using 0.1% formic acid aqueous solution and methanol as mobile phase components. The obtained coffee and adulterants extract HPLC-UV-FLD fingerprints were evaluated by partial least squares regression-discriminants analysis (PLS-DA) resulting to be excellent chemical descriptors for sample discrimination. One hundred percent classification rates for both PLS-DA calibration and prediction models were obtained. In addition, Arabica and Robusta coffee samples were adulterated with chicory, barley, and flours, and the obtained HPLC-UV-FLD fingerprints subjected to partial least squares (PLS) regression, demonstrating the feasibility of the proposed methodologies to assess coffee authenticity and to quantify adulteration levels (down to 15%), showing both calibration and prediction errors below 1.3% and 2.4%, respectively.

Keywords: coffee authenticity; HPLC-UV; HPLC-FLD; fingerprinting; chemometrics; food adulteration; chicory; barley; flours

1. Introduction

Coffee, which consists of an infusion of ground roasted beans with a characteristic taste and aroma, is among the most popular drink consumed worldwide, and has become a vital product for the economic status of the countries involved in their production and exportation. The coffee plant belongs to *Coffea* genus from the Rubiaceae family, involving more than 120 species being *Canephora coffea* (Robusta) and *Arabica coffea* (Arabica), the ones with the highest economic and commercial importance [1–4]. Coffee contains a great number of bioactive substances (like phenolic acids, polyphenols, and alkaloids; with ellagic, caffeic, and chlorogenic acids among the most abundant ones) contributing to the great properties of coffee such as its antioxidant activity, well known for its beneficial health effects. In fact, some studies have related the coffee intakes with the decrease of prevalent diseases such as cirrhosis, diabetes, cancer, and cardiovascular diseases [1,5].

Considering coffee beneficial effects and their great popularity, the market niche becomes more competitive and, consequently, the economic cut of the coffee production ends, unfortunately in many cases, in committing adulteration frauds. Coffee adulteration is mostly performed by reducing the beans quality or by adding cheaper and lower quality coffee varieties. In addition, a growing tendency is the coffee adulteration with non-coffee materials such as corn, barley, rice, chicory, middling wheat, brown sugar, soybean, rye, stems or straw, among others, to reduce cost production and increase economic benefits [3,4,6–9]. These practices are illegal and have not only economic consequences but could also imply a danger to the consumer health. Is for these reasons that food quality control of commercial coffee products to ensure coffee authenticity and to protect the consumers is very important [6,10–12]. Both targeted and untargeted analytical strategies have been described in the literature to address the discrimination, classification, and authentication of coffee samples based on the coffee region of production, their variety or their roasting degree. Some examples rely on liquid chromatography (LC) with ultraviolet (UV) [13,14] and fluorescence detection (FLD) [15], or LC [16], gas chromatography [17,18] and direct analysis in real-time (DART) [19] with mass spectrometry. However, in the last years, several works have been focused on the study of coffee adulteration cases either with coffees of inferior quality [14,15,20] or with different products such as chicory, corn, barley or wheat, among others [7–9,21–25]. For example, a targeted LC-UV method was employed by Song et al. for the quantification of six monosaccharides, trigonelline, and nicotinic acid for the identification of coffee powders adulterated with barley, wheat, and rice [8]. In another study, Cai et al. employed a targeted LC-mass spectrometry (MS) method to detect the presence of soybeans and rice in ground coffee by means of determining 17 oligosaccharides. Capillary electrophoresis coupled with mass spectrometry (CE-MS) has also been described as a targeted method for monosaccharide determination to detect coffee adulteration with soybean and corn [9].

Nowadays, untargeted fingerprinting approaches are widely employed in the literature to solve authentication problems, such as, for instance, in the case of essential oils and olive oils [26–28]. In the case of coffee, untargeted fingerprinting strategies based on nuclear magnetic resonance (NMR) [29], and laser induced breakdown (LIB) [7] spectroscopies, the use of electronic tongues [22], or digital images [23] have also been employed to detect and identify different coffee adulterations.

Based on the good performances previously demonstrated by untargeted high-performance liquid chromatography (HPLC)-UV and HPLC-FLD fingerprinting methodologies in the classification and authentication of coffees from different production regions and varieties [14,15,20], the present contribution aims at assessing the authenticity and the fraud quantitation on coffees adulterated with common adulterants such as chicory, barley, and different flours (wheat, rice, cornmeal, rye, and oatmeal). A simple liquid–solid extraction procedure based using methanol:water (50:50, v/v) was employed, and the C18 reversed-phase HPLC-UV-FLD fingerprints obtained from the analyzed methanolic aqueous extracts submitted to classificatory partial least squares regression-discriminants analysis (PLS-DA) chemometric methods to study their suitability as chemical descriptors for sample discrimination and authentication. Furthermore, PLS regression was employed as multivariate calibration method to detect and quantify adulterant levels on Arabica and Robusta coffees adulterated with chicory, barley, and flours.

2. Materials and Methods

2.1. Reagents and Chemicals

Methanol, ethanol, acetonitrile, and acetone (all of them Chromosolv™ for HPLC, ≥99.9%) were purchased from PanReac AppliChem (Barcelona, Spain). Formic acid (≥98%) was obtained from Sigma-Aldrich (St Louis, MO, USA). Water was purified with an Elix 3 coupled to a Milli-Q system from Millipore Corporation (Millipore, Bedford, MA, USA), and was filtered through a 0.22 μm nylon membrane integrated into the Milli-Q system.

2.2. Instrumentation

An Agilent 1100 Series HPLC instrument (Waldbronn, Germany) equipped with a G1312A binary pump, a WPALS G1367A automatic sample injector, a G1315B diode-array detector and a G1321A fluorescence detector connected in series, and a PC with the Agilent Chemstation software was employed to obtain the untargeted HPLC-UV and HPLC-FLD chromatographic fingerprints. Chromatographic separation was performed in a Kinetex® C18 reversed-phase (100 × 4.6 mm i.d., 2.6 µm partially porous particle size) column obtained from Phenomenex (Torrance, CA, USA). Gradient elution conditions using 0.1% formic acid in water (solvent A) and methanol (solvent B) as mobile phase components were employed. The elution program started increasing the methanol percentage from 3 to 75% in 30 min. Then, methanol increased from 75% to 95% in 2 min, and was kept at 95% methanol for 2 min more. After that, the elution program came back to the mobile phase initial conditions in 0.2 min and, finally, there was an isocratic step at 3% of methanol of 5.8 min to guarantee column re-equilibration. The injection volume was 5 µL and the mobile phase flow-rate was 0.4 mL/min. UV acquisition was performed at 280 nm and FLD acquisition at 310 nm (excitation) and at 410 nm (emission).

2.3. Samples and Sample Extraction Procedure

One hundred twenty-three samples belonging to different classes (Table 1), and purchased from supermarkets in Barcelona (Spain), Vietnam, and Cambodia, were analyzed.

Table 1. Summary of the analyzed samples.

Sample Class	Sample Type	Number of Samples
Coffee	Vietnamese Arabica coffee	13
	Vietnamese Robusta coffee	26
	Vietnamese Arabica and Robusta mixture coffee	9
	Cambodian coffee (Unknown specie)	6
Chicory	Chicory	21
Barley	Barley	6
Flour	Wheat flour	7
	Rice flour	4
	Cornmeal flour	11
	Rye flour	15
	Oatmeal flour	5

Coffee samples obtained from Vietnam were of Arabica, Robusta, and Arabica+Robusta mixture varieties. Regarding the coffee Cambodian samples, its variety was not declared in the label. Flour samples of different cereals such as wheat, rice, cornmeal, rye, and oatmeal were employed. All the analyzed samples were provided grounded by the suppliers.

Optimal sample treatment started weighing 1.00 g of sample into a 15 mL PTFE centrifuge tube (Serviquimia, Barcelona, Spain) and adding 10 mL of a methanol:water 50:50 (v/v) solution. After that, the mixture was shaken for 2 min using a Vortex (Stuart, Stone, UK). Then, the extract was centrifuged at 3500 rpm for 5 min employing a Rotanta 460 RS centrifuge (Hettich, Tuttlingen, Germany). Finally, the obtained aqueous methanolic extracts were filtered with 0.45 µm nylon filters (first mL was discarded) into an injection vial, and were stored at −4 °C until HPLC analysis. It is important to highlight that to achieve a realistic situation on coffee adulteration studies, all the barley and flour samples were submitted to a roasting process. For that purpose, 80.00 g of each sample were extended in an oven tray, and roasted for 7 min at 180 °C using a conventional oven (Teka HE 510 Me, Barcelona, Spain).

A quality control (QC) extract, prepared by mixing 50 µL of each one of the methanolic sample extracts, was used to ensure both the repeatability and robustness of the proposed methodology and the obtained chemometric results.

In addition, six coffee adulteration cases were studied involving both Vietnamese Arabica and Vietnamese Robusta coffees adulterated with chicory, barley, and wheat flour. Table 2 shows the adulteration levels (in percentage of adulterant) employed for the PLS model calibration and validation sets. An additional QC solution was also prepared at a 50% of adulterant level. For each adulteration level, five replicates were prepared, thus 55 sample extracts were analyzed for each one of the adulteration cases under study.

Table 2. Coffee and adulterant concentration levels employed for partial least squares (PLS) calibration and validation sets.

	% of Vietnamese Coffee (Arabica or Robusta)	% of Adulterant (Chicory, Barley, or Wheat Flour)
Calibration set	100	0
	80	20
	60	40
	40	60
	20	80
	0	100
Validation set	85	15
	75	25
	50	50
	25	75
	15	85

2.4. Data Analysis

Following sample treatment, the obtained methanolic extracts were randomly analyzed with the developed HPLC-UV-FLD methods. A QC and an instrumental blank (Milli-Q water) were also injected after each ten sample extracts. Different data matrices were created with the HPLC-UV or HPLC-FLD chromatographic fingerprints of the analyzed samples. The data matrices were then analyzed by partial least squares-discriminant analysis (PLS-DA) or by partial least squares (PLS) regression methods using SOLO 8.6 chemometric software obtained from Eigenvector Research (Manson, WA, USA). Description of the theoretical background of the employed chemometric methods is addressed elsewhere [30]. In any case, the X-data matrix consisted of the acquired HPLC-UV (absorbance signal vs. retention time) or HPLC-FLD (fluorescence intensity vs. retention time) chromatographic fingerprints. Instead, Y-data matrix defined each sample classes in PLS-DA, whereas defined each adulterant percentage in PLS. Chromatographic fingerprints were normalized to achieve the same weight to each variable by suppressing differences in their magnitude and amplitude scales. PLS-DA models were also validated using 70% of the samples (randomly selected) as the calibration set and the remaining 30% of the samples as the prediction set. The most appropriate number of latent variables (LVs) in PLS-DA and PLS models were established as the first significant minimum point of the cross-validation (CV) error from a Venetian blind approach.

3. Results and Discussion

3.1. Extraction Solvent Optimization

In the present contribution, untargeted HPLC-UV and HPLC-FLD fingerprints will be exploited as sample chemical descriptors to assess coffee authenticity and to quantify adulteration levels when chicory, barley, and different flours are used as coffee adulterants. Untargeted chromatographic fingerprinting strategies are based on registering instrumental signals (in this case the absorbance and the fluorescence intensity for HPLC-UV and HPLC-FLD, respectively) as a function of the retention time, but without the requirement of any information about the chemicals present in the samples, but trying to register as much instrumental discriminant signals as possible. For that purpose, simple and generic sample treatment procedures are typically applied to extract the highest number of bioactive compounds possible and belonging to different families; although, their identification or quantification is not required. With this aim, a simple liquid–solid extraction procedure was employed, and the extraction solvent composition was optimized. Different solvents such as pure water, methanol, acetonitrile, ethanol, and acetone, and the organic aqueous mixtures containing 20%, 50%, and 80% of each organic component under study (methanol, acetonitrile, ethanol, and acetone), were evaluated as extraction solvents. Four samples, a Vietnamese Arabica coffee, a Vietnamese Robusta coffee, a cornmeal flour, and a wheat flour were employed as test samples. One gram of each sample was extracted with 10 mL of each extraction solvent following the procedure described in Section 2.3, and the obtained extracts (17 different extracts for each sample under study) were analyzed with the proposed HPLC-UV and HPLC-FLD methodology following the procedure described in Section 2.2. Chromatograms with different signal profiling depending on the sample composition were obtained. The total signal area of the chemicals extracted and detected within the chromatographic segment from minute 8 to 40 was considered as chemical data for the solvent selection (the first segment of the chromatograms was not considered to remove the signal contribution from the solvents). Figure 1 shows the total signal area (normalized to the solvent extract providing the highest signal) obtained by (a) HPLC-UV and (b) HPLC-FLD for the different samples and extraction solvents evaluated. Noticeable differences were observed depending on the sample under study as well as the fingerprinting detection system; therefore, optimal conditions will be selected as a compromise of different factors. The first thing that can be observed is that pure organic solvents (methanol, acetonitrile, acetone or ethanol) extraction capacity seems to be lower in comparison to the use of organic aqueous extraction mixtures. In addition, and as a general trend, extraction capacity increases with the organic content up to a 50% and then it decreases.

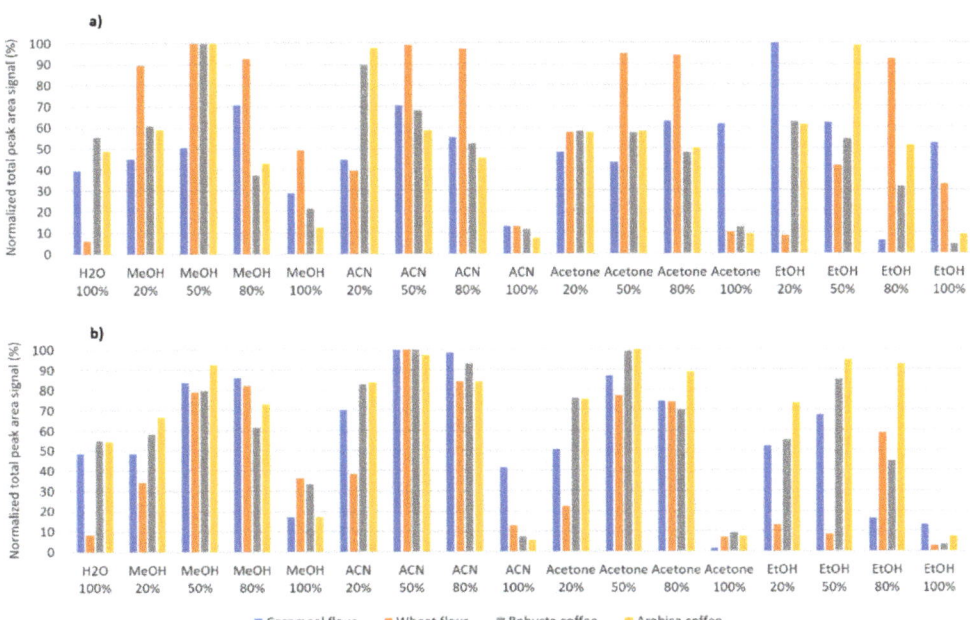

Figure 1. Total peak signal (normalized to the solvent providing the highest signal) of all the chemicals extracted with different extraction solvents and detected by (**a**) high-performance liquid chromatography with ultraviolet (HPLC-UV) and (**b**) HPLC- fluorescence detection (FLD) (within the chromatogram segment from 8 to 40 min) for a Vietnamese Arabica coffee, a Vietnamese Robusta coffee, a cornmeal flour, and a wheat flour.

The highest extraction capacity for all the samples under study when fluorescence detection is employed (Figure 1b) was achieved by using water:acetonitrile (50:50 v/v) as extraction solvent, obtaining almost the same normalized total peak area signal independently on the sample typology. In contrast, when ultraviolet detection was used (Figure 1a), better results were observed with water:methanol (50:50 v/v). In addition, this same solvent also provided a high extraction capacity with fluorescence detection, with normalized peak area signals higher than 80% for all the samples under study. Therefore, as a compromise, water:methanol (50:50 v/v) was selected as the optimal extraction solvent for the proposed liquid–solid extraction procedure. In addition, this solvent composition was more compatible to the HPLC mobile phase components.

3.2. HPLC-UV and HPLC-FLD Fingerprints

In previous works [14,15,20], we have demonstrated that HPLC-UV and HPLC-FLD fingerprints obtained after simply brewing coffees resulted in good sample chemical descriptors to address coffee classification regarding their production region, variety, and roasting degrees. This contribution aims to assess coffee authenticity when dealing with adulterations involving the use of common non-coffee-based adulterants relying on an untargeted fingerprinting strategy. For that purpose, an important number of samples belonging to different typologies (coffee, chicory, barley, and several flours) were extracted following the sample treatment previously commented, and the obtained methanolic aqueous extracts were analyzed with the proposed HPLC-UV-FLD method. For instance, Figure 2 shows the resulting HPLC-UV (a1–e1) and HPLC-FLD (a2–e2) fingerprints for randomly selected Vietnamese Arabica coffee, Vietnamese Robusta coffee, chicory, wheat flour, and barley samples. As can be seen, important differences among the number of peak signals detected as well as their relative abundances were obtained. Regarding the number of peak signals (related to the variety of sample bioactive compounds extracted), HPLC-FLD fingerprints show less signals than the HPLC-UV ones, where very few signals are detected, although comparison regarding the total abundance cannot be done. When comparing the sample typology, it is quite clear that coffee samples provide similar fingerprints independently of the detection system employed, which are completely different to those observed for the other samples. Differences related to the coffee variety (Arabica vs. Robusta) are mainly based on relative intensities of different peak signals while following a similar fingerprinting profile. This can be clearly observed, for example, on the intensity of the peak signal detected by HPLC-FLD at minute 17 for the Vietnamese Robusta coffee (Figure 2a2) which is clearly higher in comparison to the one observed in the Vietnamese Arabia coffee sample (Figure 2b2).

As commented before, the chromatographic fingerprints obtained for the samples typically employed as coffee adulterants are completely different than those observed for coffee samples, especially regarding the peak signal intensities which tend to be much lower. However, the chicory fingerprint from UV-detection (Figure 2c1) clearly disrupt with the general fingerprinting tendency obtained for the samples considered as adulterants, showing several peaks with an important signal intensity between minutes 9 and 11 in comparison to all the other samples, including the coffee ones. Regarding fluorescence fingerprints, those obtained for barley samples seem to be richer in signals detected, as well as peak intensities, in comparison to those of chicory or wheat flour. Based on these differences, and taking into consideration that fingerprints tend to be reproducible within the same sample typology, untargeted HPLC-UV and HPLC-FLD fingerprints will be evaluated as sample chemical descriptors for the characterization and classification of the analyzed samples by chemometric analysis.

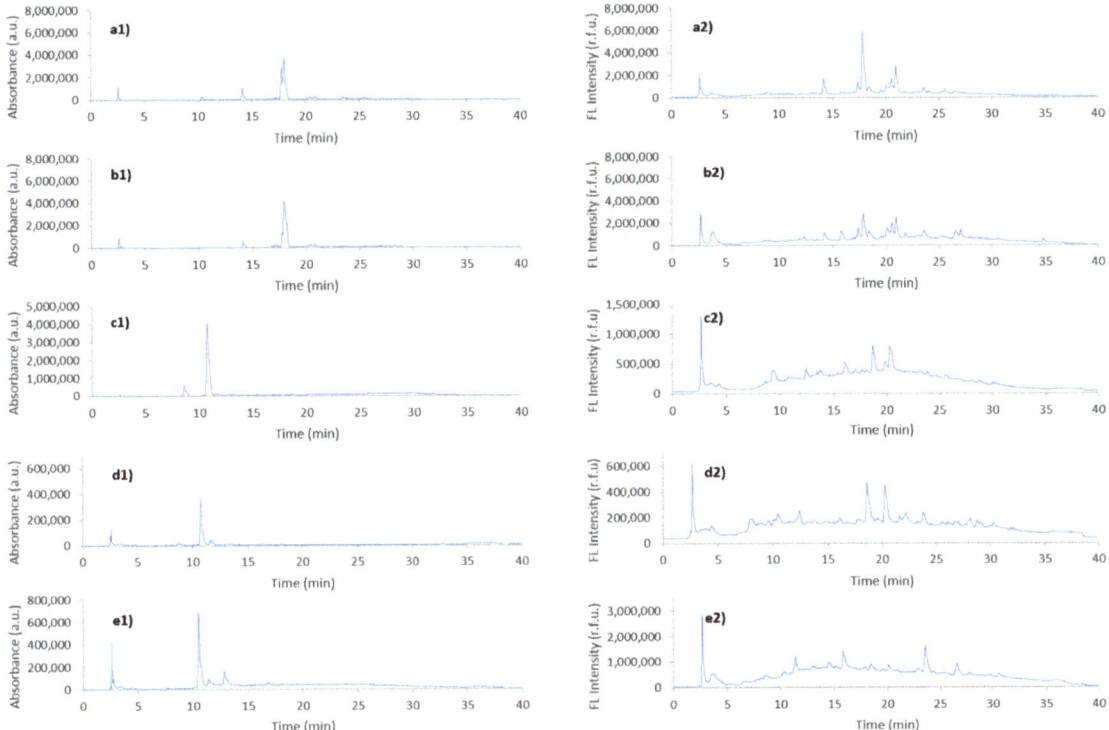

Figure 2. Untargeted HPLC-UV (a1–e1) and HPLC-FLD (a2–e2) fingerprints obtained for a selected sample of (**a**) Vietnamese Arabica coffee, (**b**) Vietnamese Robusta coffee, (**c**) chicory, (**d**) wheat flour, and (**e**) barley. UV detection was registered at 280 nm, and fluorescence detection at 310 nm (excitation) and 410 nm (emission).

3.3. Sample Characterization and Classification by Chemometrics

To evaluate if the obtained untargeted HPLC-UV and HPLC-FLD fingerprints worked properly as sample chemical descriptors for classification purposes, the methanolic extracts of 123 samples belonging to different typologies (see Table 1) were randomly analyzed, together with a QC sample which was injected every ten samples to evaluate both the reproducibility and the robustness of the proposed methodology and the obtained chemometric results. Then, the fingerprints were subjected to a classificatory PLS-DA chemometric method, and the resulting score plots defined by LV1 vs. LV2 are depicted in Figure 3. For that purpose, all the UV absorbance or the FL intensity signals, depending on the case, registered as a function of the chromatographic retention time, independently of the background noise observed, were used as data to build the chemometric matrices.

In both score plots, QCs appeared grouped in a compact cluster in the center area of the plot, which ensures the reproducibility of the proposed HPLC fingerprinting methodology as well as the robustness of the chemometric results. In addition, samples tend to be well grouped according to their typology, with the exception of chicory samples which form a more disperse group although perfectly discriminated from the other sample types, which may be related to the different brand and roasting process. Flour samples also appeared in quite a compacted group independently of the type or cereal (wheat, rice, cornmeal, rye, and oatmeal). Sample distribution within the score plots depends on the HPLC fingerprints used as chemical descriptors. Thus, when HPLC-UV fingerprints are employed (Figure 3a) coffee samples tend to exhibit negative LV2 values, while adulterants show positive LV2 values, and are separated from flours, barley to chicory sample with the increase in LV1 values. As a result, the four groups of samples under study are perfectly discriminated. In contrast, with HPLC-FLD fingerprints, full discrimination of all the sample groups was not accomplished. Coffee samples exhibited positive and negative LV2 and LV1 values, respectively, and are partially overlapped with barley samples; although, this last group tend to be exhibiting mainly negative LV2 values. In any case, discrimination between the three groups of adulterant samples was also accomplished, but both LV1 and LV2 are playing an important role.

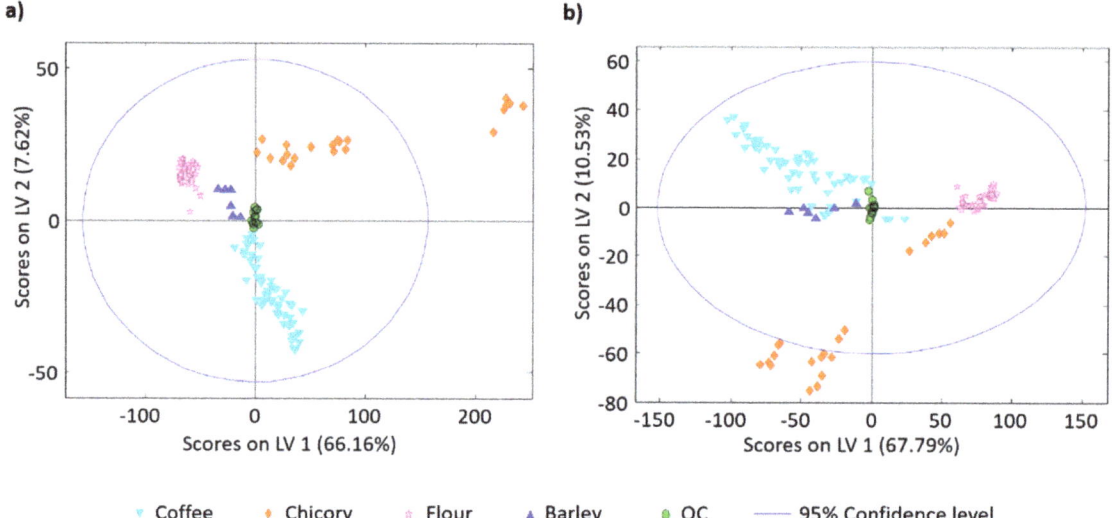

Figure 3. Partial least squares regression-discriminants analysis (PLS-DA) score plots of LV1 vs. LV2 for the classification of the analyzed samples when untargeted (**a**) HPLC-UV and (**b**) HPLC-FLD fingerprints were employed as sample chemical descriptors. PLS-DA models were built with 2 and 3 LVs for HPLC-UV and HPLC-FLD, respectively.

As previously commented, the present contribution aims to assess coffee authenticity when adulterations with chicory, barley, or flours are taking place. For that purpose, PLS-DA models of coffee against each one of the adulterants were validated to determine the model classification rate. Thus, paired PLS-DA models were built using 70% of the samples of each group, randomly selected, as the calibration set, and the remaining 30% of samples as a validation set. They were considered as unknown samples for prediction purposes in order to evaluate the model classification performances. Figure 4 shows the obtained results for the paired PLS-DA model validations when (1) HPLC-UV and (2) HPLC-FLD fingerprints were employed as sample chemical descriptors for the classification studies of coffee against chicory (Figure 4(a1,2)), flour (Figure 4(b1,2)), and barley (Figure 4(c1,2)) adulterants. As can be seen, 100% classification rates for calibration and validation were obtained using both HPLC-UV and HPLC-FLD fingerprinting methodologies, demonstrating the feasibility of the proposed untargeted fingerprinting strategy to assess coffee classification and authentication against common non-based coffee adulterants such as chicory, barley, and flours from different cereals.

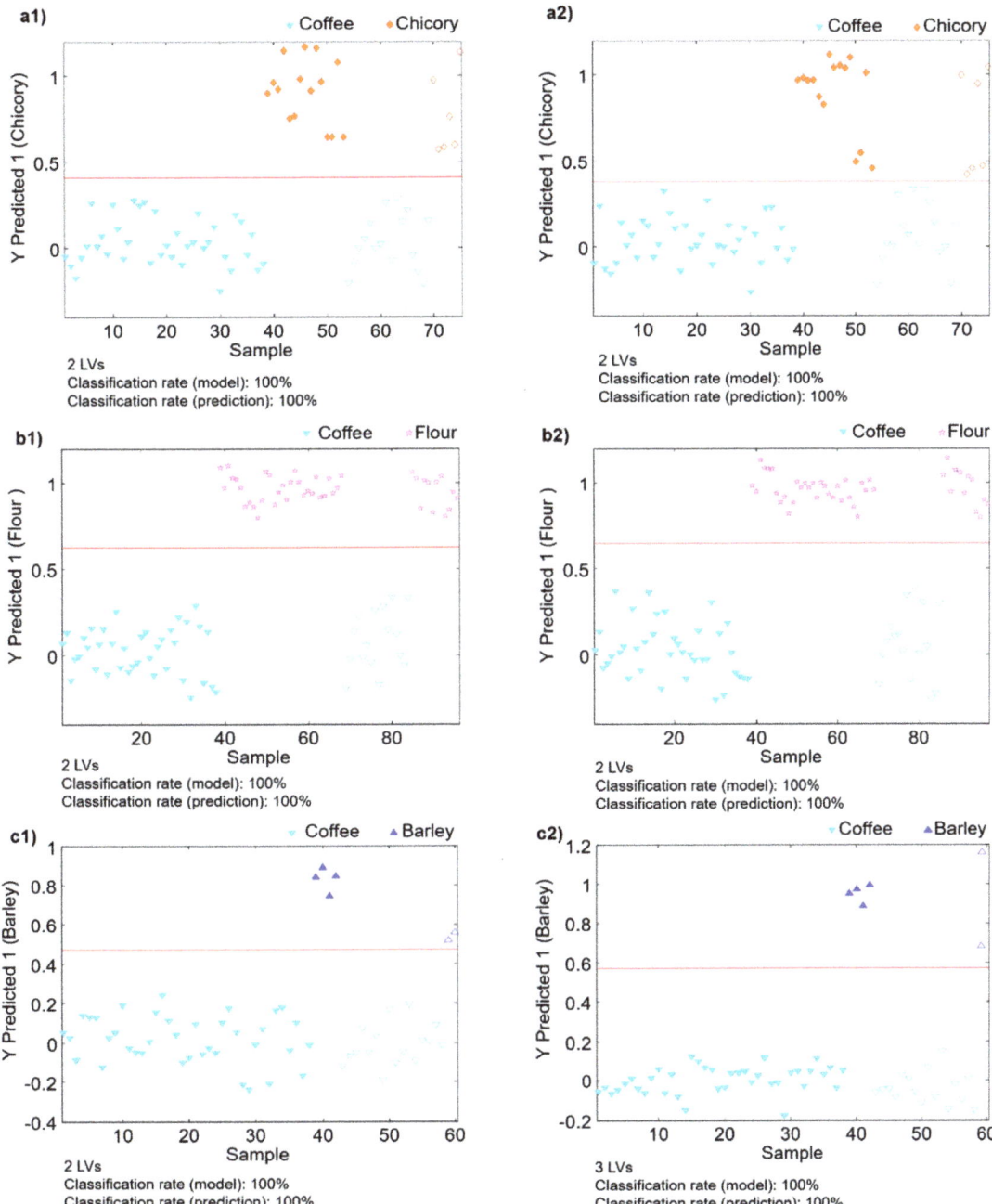

Figure 4. Classification plots defined by the sample vs. the predicted classes when (1) HPLC-UV (2) HPLC-FLD fingerprints were used as sample chemical descriptors. (**a**) Coffee vs. chicory samples, (**b**) coffee vs. flour samples, and (**c**) coffee vs. barley samples. Filled symbols correspond to the calibration set and empty symbols correspond to the validation set (unknown samples for prediction purposes).

3.4. Quantitation of Adulteration Levels by PLS

The capacity of the untargeted HPLC-UV and HPLC-FLD fingerprinting methodologies to detect frauds and to quantify coffee adulteration levels was evaluated by PLS regression studying six adulterations cases based on both Vietnamese Arabica and Robusta coffees, each one adulterated with chicory, barley, and wheat flour, respectively. For each adulteration case under study, two independent sets of samples with different adulterant concentration levels were prepared for calibration and validation purposes, as described in Table 2. The samples were then extracted using the proposed sample treatment procedure, and the obtained methanolic aqueous extracts were randomly analyzed with the untargeted HPLC-UV-FLD method. The obtained chromatographic fingerprints were then employed as sample chemical descriptors and submitted to PLS for quantitation purposes. As an example, Figure 5 shows the scatter plots of Y measured vs. Y predicted obtained for adulteration of the Vietnamese Arabica coffee with a wheat flour when (a) HPLC-UV and (b) HPLC-FLD fingerprints were used as sample chemical descriptors.

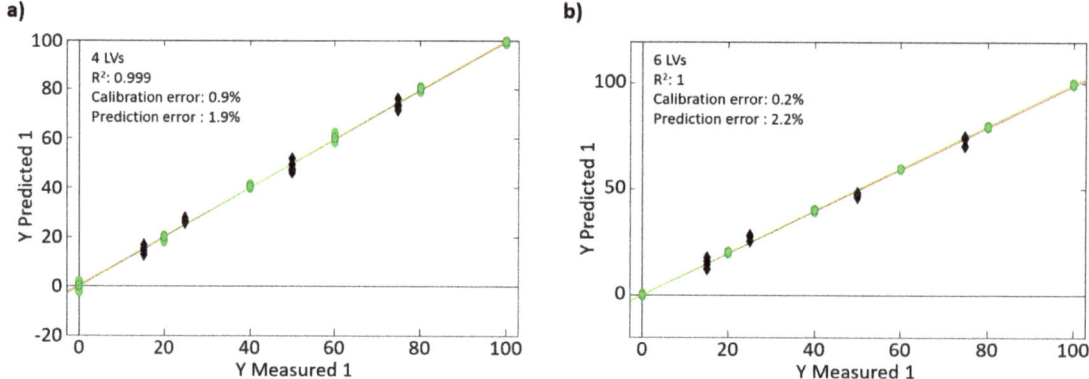

Figure 5. PLS regression scatter plots of measured vs. predicted percentages of adulterant for the adulteration case of Vietnamese Arabica coffee with a wheat flour when (**a**) HPLC-UV and (**b**) HPLC-FLD fingerprints were used as sample chemical descriptors.

The statistic PLS regression parameters obtained with the six adulteration cases under study and the number of LVs to build the PLS models are summarized in Table 3. As can be seen, very good results were obtained, with calibration and prediction errors always below of 1.4% and 2.4%, respectively. Both, untargeted HPLC-UV and HPLC-FLD fingerprints seem to be appropriate sample chemical descriptors for the fraud detection and quantitation, resulting in similar calibration errors (0.2–1.4% with UV and 0.2–1.3% with FLD) and prediction errors (0.9–2.2% with UV and 0.4–2.4% with FLD).

It should be highlighted that these results are much better than those obtained when HPLC-UV and HPLC-FLD were used as sample chemical descriptors to detect and quantify coffee frauds based on adulteration with coffees of different production regions and different varieties, were calibration errors up to 3.4% and 2.9% were reported for UV and FLD, respectively, and prediction errors up to 7.5% and 18.3%, respectively [14,20]. This is probably due to the higher differences found in the chromatographic fingerprints among coffees and adulterants.

Table 3. PLS results for the six adulteration cases studied based on Vietnamese Arabica and Vietnamese Robusta coffees adulterated with chicory, wheat, flour, and barley.

Method	Adulterant	PLS Parameter	Vietnamese Arabica Coffee	Vietnamese Robusta Coffee
HPLC-UV fingerprinting	Chicory	LVs	5	4
		Calibration error (%)	0.2	0.6
		Prediction error (%)	1.2	0.9
	Wheat Flour	LVs	4	4
		Calibration error (%)	0.9	0.4
		Prediction error (%)	1.9	1.5
	Barley	LVs	3	3
		Calibration error (%)	1.4	1.0
		Prediction error (%)	1.5	2.2
HPLC-FLD fingerprinting	Chicory	LVs	4	3
		Calibration error (%)	0.5	0.9
		Prediction error (%)	1.1	2.0
	Wheat Flour	LVs	6	4
		Calibration error (%)	0.2	0.3
		Prediction error (%)	2.2	1.0
	Barley	LVs	4	6
		Calibration error (%)	0.4	1.3
		Prediction error (%)	0.4	2.4

These results demonstrate the feasibility of both untargeted HPLC-UV and HPLC-FLD fingerprints of methanolic sample extracts as good sample chemical descriptors to address the detection and quantitation of adulterant levels in fraudulent coffee samples adulterated with non-based coffee adulterants such as chicory, barley, and flour.

4. Conclusions

Both untargeted HPLC-UV and HPLC-FLD fingerprints obtained after a sample extraction using water:methanol (50:50 v/v) have proved to be suitable sample chemical descriptors to assess the classification and authentication of coffee samples in front of common coffee adulterants such as chicory, barley, and flours. Excellent discrimination of coffee samples and the proposed adulterants was achieved by exploratory PLS-DA, especially when using HPLC-UV fingerprints. Moreover, 100% sample classification rates for both calibration and prediction were obtained when validating paired PLS-DA models of either Vietnamese Arabica or Robusta coffee against each one of the studied adulterants (chicory, barley, and flour) demonstrating the classification and authentication capacity of the proposed methodology.

Finally, PLS multivariate calibration was applied to six adulteration cases involving a Vietnamese Robusta and a Vietnamese Arabica coffees adulterated at different levels with chicory, barley, and wheat flour, and the proposed untargeted HPLC-UV and HPLC-FLD fingerprints were appropriate to detect and quantify the adulterant levels down to 15% (lowest level evaluated for prediction) with good calibration and prediction errors (values always lower than 1.3% and 2.4%, respectively).

The proposed untargeted HPLC-UV and HPLC-FLD fingerprinting methods can be used as a simple, reliable, and relatively economic approach to assess and guarantee coffee authenticity, and to prevent fraudulent practices against adulteration with common non-coffee-based adulterants such as chicory, barley, and flours. The simplicity of an untargeted fingerprinting approach, without the requirement of using chemical standards to quantify targeted compounds, makes this methodology ideal to prevent frauds in developing coffee production countries.

Author Contributions: Conceptualization, J.S. and O.N.; methodology, N.N.; validation, N.N.; formal analysis, N.N.; investigation, N.N., J.S., and O.N.; writing—original draft preparation, N.N.;

writing—review and editing, N.N., J.S., and O.N.; supervision, J.S. and O.N.; funding acquisition, J.S. and O.N. All authors have read and agreed to the published version of the manuscript.

Funding: This research was supported by the Agency for Administration of University and Research Grants (Generalitat de Catalunya, Spain) under the projects 2017SGR-171 and 2017SGR-310.

Data Availability Statement: Data is available upon request to the authors.

Conflicts of Interest: The authors declare no conflict of interest. The funders had no role in the design of the study; in the collection, analyses, or interpretation of data; in the writing of the manuscript, or in the decision to publish the results.

References

1. Esquivel, P.; Jiménez, V.M. Functional properties of coffee and coffee by-products. *Food Res. Int.* **2012**, *46*, 488–495. [CrossRef]
2. Naranjo, M.; Vélez, I.L.T.; Benjamín, I.I.; Iii, A.R. Actividad antioxidante de café colombiano de diferentes calidades Antioxidant activity of different grades of Colombian coffee. *Rev. Cub. Plant Med.* **2011**, *16*, 164–173.
3. Thornburn Burns, D.T.; Tweed, L.; Walker, M.J. Ground Roast Coffee: Review of Analytical Strategies to Estimate Geographic Origin, Species Authenticity and Adulteration by Dilution. *Food Anal. Methods* **2017**, *10*, 2302–2310. [CrossRef]
4. Toci, A.T.; Farah, A.; Pezza, H.R.; Pezza, L. Coffee Adulteration: More than Two Decades of Research. *Crit. Rev. Anal. Chem.* **2015**, *46*, 83–92. [CrossRef] [PubMed]
5. Crozier, A.; Ahihara, H.; Tomás-Barbéran, F. (Eds.) *Teas, Cocoa and Coffee. Plant Secondary Metabolites and Health*; Wiley-Blackwell: Oxford, UK, 2012; ISBN 9781444334418.
6. Kamiloglu, S. Authenticity and traceability in beverages. *Food Chem.* **2019**, *277*, 12–24. [CrossRef]
7. Sezer, B.; Apaydin, H.; Bilge, G.; Boyaci, I.H. Coffee arabica adulteration: Detection of wheat, corn and chickpea. *Food Chem.* **2018**, *264*, 142–148. [CrossRef]
8. Song, H.Y.; Jang, H.W.; Debnath, T.; Lee, K.-G. Analytical method to detect adulteration of ground roasted coffee. *Int. J. Food Sci. Technol.* **2018**, *54*, 256–262. [CrossRef]
9. Daniel, D.; Lopes, F.S.; dos Santos, V.B.; do Lago, C.L. Detection of coffee adulteration with soybean and corn by capillary electrophoresis-tandem mass spectrometry. *Food Chem.* **2018**, *243*, 305–310. [CrossRef] [PubMed]
10. Campmajó, G.; Núñez, N.; Núñez, O. The Role of Liquid Chromatography-Mass Spectrometry in Food Integrity and Authenticity. In *Mass Spectrometry—Future Perceptions and Applications*; Kamble, G.S., Ed.; IntechOpen: London, UK, 2019; pp. 3–20.
11. Moore, J.C.; Spink, J.; Lipp, M. Development and Application of a Database of Food Ingredient Fraud and Economically Motivated Adulteration from 1980 to 2010. *J. Food Sci.* **2012**, *77*, R118–R126. [CrossRef]
12. Gonzalvez, A.; Armenta, S.; Guardia, M. De Trace-element composition and stable-isotope ratio for discrimination of foods with Protected Designation of Origin. *Trends Anal. Chem.* **2009**, *28*, 1295–1311. [CrossRef]
13. De Luca, S.; De Filippis, M.; Bucci, R.; Magrì, A.D.; Magrì, A.L.; Marini, F. Characterization of the effects of different roasting conditions on coffee samples of different geographical origins by HPLC-DAD, NIR and chemometrics. *Microchem. J.* **2016**, *129*, 348–361. [CrossRef]
14. Núñez, N.; Collado, X.; Martínez, C.; Saurina, J.; Núñez, O. Authentication of the Origin, Variety and Roasting Degree of Coffee Samples by Non-Targeted HPLC-UV Fingerprinting and Chemometrics. Application to the Detection and Quantitation of Adulterated Coffee Samples. *Foods* **2020**, *9*, 378. [CrossRef]
15. Núñez, N.; Martínez, C.; Saurina, J.; Núñez, O. High-performance liquid chromatography with fluorescence detection fingerprints as chemical descriptors to authenticate the origin, variety and roasting degree of coffee by multivariate chemometric methods. *J. Sci. Food Agric.* **2021**, *101*, 65–73. [CrossRef] [PubMed]
16. Pérez-Míguez, R.; Sánchez-López, E.; Plaza, M.; Marina, M.L.; Castro-Puyana, M. Capillary electrophoresis-mass spectrometry metabolic fingerprinting of green and roasted coffee. *J. Chromatogr. A* **2019**, *1605*, 360353. [CrossRef] [PubMed]
17. Mehari, B.; Redi-Abshiro, M.; Chandravanshi, B.S.; Combrinck, S.; McCrindle, R.; Atlabachew, M. GC-MS profiling of fatty acids in green coffee (Coffea arabica L.) beans and chemometric modeling for tracing geographical origins from Ethiopia. *J. Sci. Food Agric.* **2019**, *99*, 3811–3823. [CrossRef] [PubMed]
18. Ongo, E.A.; Montevecchi, G.; Antonelli, A.; Sbervegleri, V.; Sevilla, F. Metabolomics fingerprint of Philippine coffee by SPME-GC-MS for geographical and varietal classification. *Food Res. Int.* **2020**, *134*, 109227. [CrossRef] [PubMed]
19. Danhelova, H.; Hradecky, J.; Prinosilova, S.; Cajka, T.; Riddellova, K.; Vaclavik, L.; Hajslova, J. Rapid analysis of caffeine in various coffee samples employing direct analysis in real-time ionization–high-resolution mass spectrometry. *Anal. Bioanal. Chem.* **2012**, *403*, 2883–2889. [CrossRef]
20. Núñez, N.; Saurina, J.; Núñez, O. Non-targeted HPLC-FLD fingerprinting for the detection and quantitation of adulterated coffee samples by chemometrics. *Food Control.* **2021**, *124*, 107912. [CrossRef]
21. Cai, T.; Ting, H.; Jin-Lan, Z. Novel identification strategy for ground coffee adulteration based on UPLC–HRMS oligosaccharide profiling. *Food Chem.* **2016**, *190*, 1046–1049. [CrossRef]
22. De Morais, T.C.B.; Rodrigues, D.R.; de Carvalho Polari Souto, U.T.; Lemos, S.G. A simple voltammetric electronic tongue for the analysis of coffee adulterations. *Food Chem.* **2019**, *273*, 31–38. [CrossRef]

23. Souto, U.T.D.C.P.; Barbosa, M.F.; Dantas, H.V.; De Pontes, A.S.; Lyra, W.D.S.; Diniz, P.H.G.D.; De Araújo, M.C.U.; Da Silva, E.C. Screening for Coffee Adulteration Using Digital Images and SPA-LDA. *Food Anal. Methods* **2014**, *8*, 1515–1521. [CrossRef]
24. Reis, N.; Botelho, B.G.; Franca, A.S.; Oliveira, L.S. Simultaneous Detection of Multiple Adulterants in Ground Roasted Coffee by ATR-FTIR Spectroscopy and Data Fusion. *Food Anal. Methods* **2017**, *10*, 2700–2709. [CrossRef]
25. Winkler-Moser, J.K.; Singh, M.; Rennick, K.A.; Bakota, E.L.; Jham, G.N.; Liu, S.X.; Vaughn, S.F. Detection of Corn Adulteration in Brazilian Coffee (Coffea arabica) by Tocopherol Profiling and Near-Infrared (NIR) Spectroscopy. *J. Agric. Food Chem.* **2015**, *63*, 10662–10668. [CrossRef] [PubMed]
26. Taghadomi-Saberi, S.; Garcia, S.M.; Masoumi, A.A.; Sadeghi, M.; Marco, S. Classification of Bitter Orange Essential Oils According to Fruit Ripening Stage by Untargeted Chemical Profiling and Machine Learning. *Sensors* **2018**, *18*, 1922. [CrossRef] [PubMed]
27. Beale, D.J.; Morrison, P.D.; Karpe, A.V.; Dunn, M.S. Chemometric Analysis of Lavender Essential Oils Using Targeted and Untargeted GC-MS Acquired Data for the Rapid Identification and Characterization of Oil Quality. *Molecules* **2017**, *22*, 1339. [CrossRef] [PubMed]
28. Barbieri, S.; Cevoli, C.; Bendini, A.; Quintanilla-Casas, B.; García-González, D.L.; Toschi, T.G. Flash Gas Chromatography in Tandem with Chemometrics: A Rapid Screening Tool for Quality Grades of Virgin Olive Oils. *Foods* **2020**, *9*, 862. [CrossRef]
29. Milani, M.I.; Rossini, E.L.; Catelani, T.A.; Pezza, L.; Toci, A.T.; Pezza, H.R. Authentication of roasted and ground coffee samples containing multiple adulterants using NMR and a chemometric approach. *Food Control.* **2020**, *112*, 107104. [CrossRef]
30. Massart, D.L.; Vandeginste, B.G.M.; Buydens, L.M.C.; de Jong, S.; Lewi, P.J.; Smeyers-Verbeke, J. *Handbook of Chemometrics and Qualimetrics*; Elsevier: Amsterdam, The Netherlands, 1997.

Communication

Alkaline Phosphatase Survey in Pecorino Siciliano PDO Cheese

Massimo Todaro [1,*], Vittorio Lo Presti [2], Alessandro Macaluso [3], Maria Alleri [4], Giuseppe Licitra [5] and Vincenzo Chiofalo [3]

1. Dipartimento Scienze Agrarie Alimentari Forestali, University of Palermo, viale delle Scienze, 13, 90128 Palermo, Italy
2. Dipartimento di Scienze Veterinarie, University of Messina, viale dell'Annunziata, 98168 Messina, Italy; vittorio.lopresti@unime.it
3. Dipartimento di Scienze Chimiche, Biologiche, Farmaceutiche ed Ambientali, University of Messina, viale dell'Annunziata, 98166 Messina, Italy; alex.macaluso@virgilio.it (A.M.); vincenzo.chiofalo@unime.it (V.C.)
4. Nuovo Consorzio di Tutela del Pecorino Siciliano DOP, via dell'amicizia 26, 91020 Poggioreale (TP), Italy; maria.alleri@unipa.it
5. Consorzio per la Ricerca nel Settore della Filiera Lattiero-Casearia e dell'Agroalimentare, SP 25, 97100 Ragusa, Italy; glicitra@unict.it
* Correspondence: massimo.todaro@unipa.it

Abstract: The determination of alkaline phosphatase (ALP) in cheeses has become an official method for controlling cheeses with a protected designation of origin (PDO), all of which use raw milk. PDO cheeses, characterized by high craftsmanship, usually have an uneven quality. However, for these cheeses, it is necessary to establish ALP values so that they can be defined as a raw milk product. In this study, a dataset with Pecorino Siciliano PDO samples was analyzed to determine ALP both at the core and under the rind. The results showed that there was no significant difference between the different zones in Pecorino cheese. A second dataset of 100 pecorino cheese samples determined that ALP was only at the core of the cheese. Moreover, there was a statistically significant difference between the ALP values of cheeses produced with raw milk and those produced with pasteurized milk. Furthermore, according to the temperatures, a wide variability of ALP values was observed in the Pecorino Siciliano PDO samples from the core of the cheeses. This was a result of several under scotta whey cooking methodologies adopted by cheesemakers, which do not permit a clear range. Therefore, further investigation is desirable.

Keywords: alkaline phosphatase determination; PDO Pecorino Siciliano cheese; raw milk determination

1. Introduction

Italy has a long history of cheese production. In Sicily, the largest island in the Mediterranean area, the Phoenician community began the production of cheese. Several archaeological finds have indicated that dairy activity was routinely conducted during the Eneolithic age [1]. Therefore, Pecorino Siciliano PDO is considered the oldest cheese in the EU.

Pecorino Siciliano is a traditional Italian PDO cheese produced throughout Sicily. It is a semi-hard cheese that is manufactured using traditional techniques, i.e., from raw ewe's milk without any bacterial starters, according to production protocol (GUCE C 170 EUR-Lex-52020XC0518(03)) (Figure 1). Artisanal cheese-making that uses traditional wooden equipment causes a microbiota that is responsible for acidifying curd and maturing cheese that originates from raw milk. This impacts the equipment, the animal rennet, and the transformation of the dairy environment [2]. PDO Pecorino Siciliano cheese is defined as a semi-cooked cheese because the cheese is cooked under hot scotta whey. Scotta whey is a residual product created during the extraction of ricotta cheese; normally, it is used for cooking pecorino cheeses at 74–78 °C for at least two to three hours. The production protocol does not consider where cooking takes place; therefore, some producers use wooden vats and others use steel. Moreover, some producers use either steel or copper boilers. It is well known that the size and type of the container used to cook cheese influences the degree of heat penetration, as does the amount of scotta whey used and its temperature.

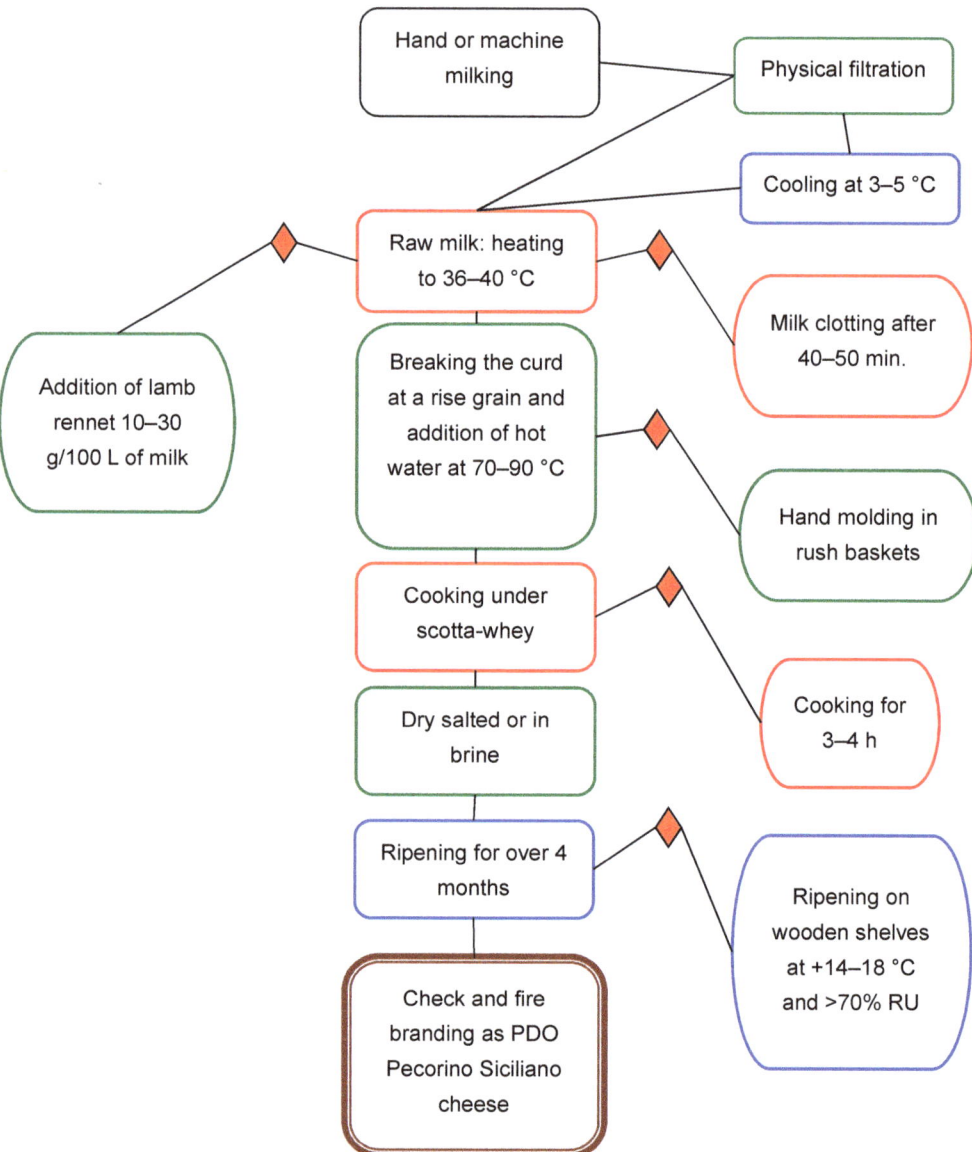

Figure 1. Flow chart of PDO Pecorino Siciliano cheese.

To safeguard PDO production from food fraud, it is necessary to develop a control system for raw milk when producing this type of cheese. To date, one of the most widely used analytical systems is the determination of alkaline phosphatase activity (ALP).

ALP is used throughout the world as a marker for the proper pasteurization of milk, because it guarantees hygienic safety [3]. The analysis of ALP in cheese has been described by ISO 11816-2/IDF 155-2 [4]. In the past, this proposed method was not considered appropriate for reflecting the heat treatment of cheese milk in some types of cheese [5]. For cheese, no legal limit has been set, as has been legislated for milk [6]. Further studies, however, have evidenced that processing conditions, texture, size, and high variability can impact the residual activity of ALP and its zonal distribution in cheese [7–9]. The temperature at which the curd is heated, as well as the size of the cheese wheel, are the main parameters that influence the residual ALP activity in cheese. Therefore, as well as having a reliable analytical method, there is an urgent need for an appropriate limit for residual ALP activity to characterize cheeses made from pasteurized milk. Based on a study that involved 700 cheese samples from 32 different cheese varieties, Egger et al. [10] proposed a limit for ALP activity (10 mU/g) in cheese developed from pasteurized milk.

Pecorino Siciliano PDO cheese, similar to several artisanal pecorino cheeses produced in southern Italy, is characterized by a wide variability [2,11]. Therefore, at the request of the Protection Consortium, a survey was conducted on Pecorino Siciliano cheese samples subjected to PDO certification in order to define ALP values.

2. Materials and Methods

2.1. Cheese Production and Sampling

Two datasets of Pecorino Siciliano cheese samples were used in this survey:

(1) A total of 98 Pecorino Siciliano cheese samples (0.5 kg), 5–month ripened, were taken. In total, 78 came from 9 dairies in the PDO Protection Consortium. The remaining 20 samples (0.5 kg) were produced in Sicily and came from 10 dairies that declared, on their labels, that the product was "produced with pasteurized milk". The ALP analysis was detected at the core of the cheese;

(2) A total of 36 Pecorino Siciliano cheese samples (0.5 kg), 5–month ripened, came from 6 dairies in the PDO Protection Consortium and from 3 dairies that declared, on their labels, that the product was "produced with pasteurized milk". The ALP analysis was detected at the core and under the rind of the cheese.

PDO Pecorino Siciliano cheese is cooked under scotta whey, according to the approved protocol (Figure 1). Then, producers choose cooking temperatures, vats, and cooking time on the basis of environmental conditions and their cheese-making experience. This cooking system determines a wide variability in cheese quality and composition. To better understand the results of the conducted survey, cheese temperatures and cooking technology during PDO Pecorino Siciliano production were detected in 9 dairies that belonged to the PDO Protection Consortium (Table 1). Depending on the cooking processes and the temperature detected at the core after cooking stopped, dairies were classified as either weak (t < 47 °C) or severe (t ≥ 47 °C). Moreover, a further class was formed and called "mixed", which included cheeses produced by cheesemakers who apply a severe cooking method and lower cooking temperatures (Table 1).

Table 1. Cooking technology detected in dairies belonging to the PDO Protection Consortium for Pecorino Siciliano cheese.

Dairy	Temperature of Cheese before Cooking (°C)	Temperature of Scotta Whey before Cooking (°C)	Ratio between Scotta Whey Liters and kg of Cooked Cheese (L/kg)	Temperature of Cheese after Cooking (°C)	Temperature of Scotta Whey after Cooking (°C)	Cooking Time (h)	Cooking Classes
A	39.5	75.0	5	54.5	57.5	4.5	severe
B	43.0	74.0	4	52.0	54.0	4	severe
C	39.7	74.0	3.5	50.0	59.0	4.5	severe
D	42.0	77.0	4	47.0	58.0	3	severe
E	38.5	78.0	3.5	47.5	65.0	3.5	severe
F	38.8	72.5	3	49.5	59.2	2	severe
H	34,5	73.5	2	44.7	52.2	3	weak
I	41.0	77.0	3	44.0	44.0	3	weak
L	37.5	75.0	3	45.5	52.0	3	weak
Other PDO cheesemaker	38.0	74.7	3	45.0	34.5	2.8	mixed

2.2. Analysis of Alkaline Phosphatase with Fluorophos®

Cheese samples were planned by the PDO Protection Consortium and the Corfilac Consortium. From each, a slice of cheese (0.5 kg) was taken, vacuum-packed, and transferred to a laboratory at the University of Messina. Samples were refrigerated at $5 \pm 2\,°C$. The samples were weighed, and the size of each slice was measured.

All samples were analyzed at the core. Altogether, 98 cheeses samples were taken from the central part of the slice (core), and the distance from the core to the surface was measured (Figure 2). The size of the Pecorino wheels, about 7 kg in weight, gave a height of 17 cm and a diameter of 25 cm.

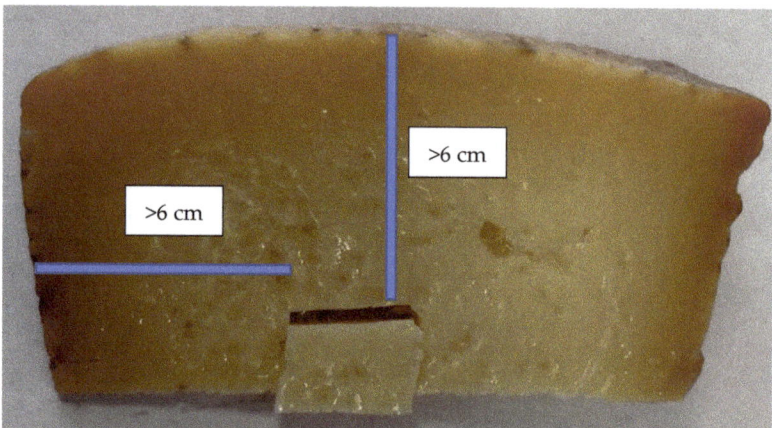

Figure 2. Cheese sample taken at core during the ALP analysis.

ALP analyses were conducted at the core and under the rind in 9 cheeses, 6 of which were made with raw milk and 3 of which were made with pasteurized milk. In total, 98 cheeses were sampled. For each cheese, a 1 cm × 1 cm section was taken: 2 at the core and 2 under the rind. The distance of each sample taken at core was higher than 6 cm from the core to the surface, while distance of each sample taken under the rind was approximately 1 cm under the rind (Figure 3). Each portion was finely ground and analyzed, as described below. Cheese samples were analyzed according to the ISO11816-2/IDF 155-2 (second edition 2016.08.15), apart from some optimizations.

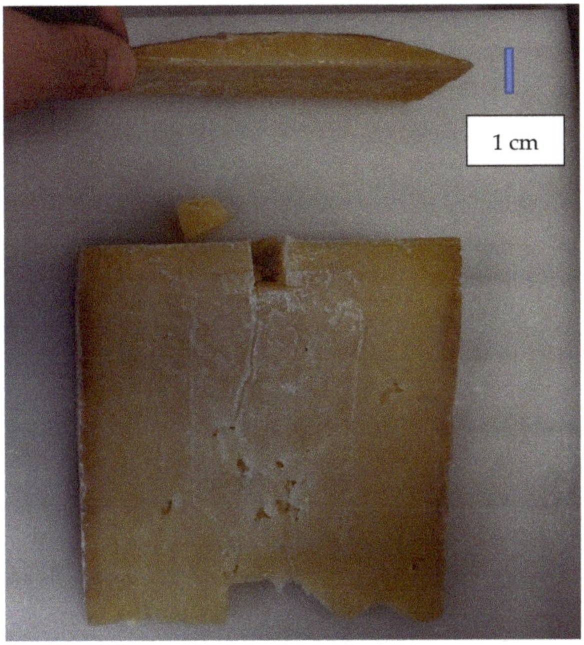

Figure 3. Cheese sample taken under the rind during the ALP analysis.

Analyses were carried out using a Fluorophos® ALP test system (Advanced Instruments Inc., Norwood, MA, USA) and calibrated with a Calibrator Set FLA250 (Fluorophos® Advanced Instruments Inc., Norwood MA, USA) using a pasteurized pecorino. This was undertaken in order to consider the matrix effect.

A cheese extraction buffer FLA005 (Fluorophos® Advanced Instruments Inc., Norwood MA, USA) was used to extract the cheese. An aliquot in the cheese grinder was weighed to the nearest 1 mg in a 15 mL conical test tube. Next, 5 mL of the cheese buffer was added and mixed with a homogenizer (Ultra-Turrax® T 25 basic IKA®-WERKE, Janke and Kunkel-Str. 10 79219 Staufen, Germany) for 60 s to obtain a completely homogenous dispersion. Another 5 mL Cheese Extraction Buffer was used to rinse the homogenizer and this emulsion was added to the previous extract.

The final sample was centrifuged at 1000 g/min at 4 °C for 10 min. The upper phase was collected in a clean tube, wherein 25–75 µL were withdrawn for analysis, according to ISO 11816-2/IDF 155-2 [4].

The instrumental results were converted into dilution factors and expressed in mU/g. Before each analysis, the instrument was checked according to the manufacturer's instructions and the ISO ISO11816-2/IDF 155-2. This was accomplished by performing tests with a Daily Instrument Control® FLA 280 and Phospha Check Pasteurization Control® FLA260 (Fluorophos® Advanced Instruments Inc., Norwood, MA, USA).

2.3. Statistical Analysis

2.3.1. First Data Set

Here, a total of 36 cheese samples were statistically analyzed. We detected a lack of normal distribution for ALP values. As such, a logarithm transformation of the original data was applied. The effect that the type of sample (core or under-rind) had on the analysis was detected using ANOVA analysis. Using a GLM procedure [12], the following model was implemented:

$$ALP_{jlk} = \mu + Sample_j + Milk_l + (Sample \times Milk)_{jl} + \varepsilon_{jlk} \quad (1)$$

where ALP_{ijk} is the ALP logarithm value; $Sample_j$ is the part of the cheese taken, wherein j indicates the core or under the rind; $Milk_j$ is the type of milk used for cheesemaking, wherein j indicates either raw or pasteurized. Means were compared using the Student's *t*-test. Moreover, *p* values less than 0.05 were considered to be statistically significant.

2.3.2. Second Data Set

In this study, at least 98 cheese samples were analyzed, and simple statistics were calculated. We detected a lack of a normal distribution for ALP values. As such, we applied a logarithm transformation to the original data. The effect of under scotta whey cheese cooking on ALP logarithmic values was detected using ANOVA analysis. Using a GLM procedure [12], the following model was implemented:

$$ALP_{jk} = \mu + Cooking_j + \varepsilon_{jk} \quad (2)$$

where ALP_{ijk} is the ALP logarithm value and $Cooking_j$ is the classification of under scotta whey cheese cooking, wherein j indicates severe, weak, mixed, or pasteurized milk. Means were compared using the Student's *t*-test. Moreover, *p* values less than 0.05 were considered to be statistically significant.

3. Results and Discussion

Several studies reported that cheese size impacts the residual activity of ALP and its zonal distribution [6–10]. Additionally, several authors agree that the temperature at which the curd is heated, and the size of the cheese wheel, are the main parameters that influence residual ALP activity in cheese [6,10].

The particular cooking technology applied to PDO Pecorino Siciliano cheese requires in-depth study in order to determine ALP values in different areas of the cheese. Table 2 shows the statistical analyses of 36 Pecorino Siciliano cheese samples. As expected, ALP values detected on cheeses produced with raw milk showed significantly higher values than cheeses produced with pasteurized milk. The effect of the cheese area was not statistically significant for PDO cheeses produced with raw milk, while significant differences in ALP values were found between the core and under-rind areas in cheeses produced with pasteurized milk. In accordance with Egger et al. [10], ALP values detected at the core were significantly lower than ALP values under the rind, probably due to a higher temperature at the core of the cheese and slower cooling. The cheese area (core or under-rind) in PDO Pecorino Siciliano cheeses was found to have no significant influence on the ALP value, probably due to the wide variability that characterizes these cheeses. This is because they are produced in an artisanal way, wherein the cheesemaker is able to modify the quality of the cheese while respecting PDO protocol. The effect of the cooking process (i.e., severe, lack, and mixed) was tested in the statistical model, but it was not significant and, thus, was deleted.

Table 2. ALP values in different zones of the cheese slice.

Milk	Sample	ALP (log 10)	Standard Error	Probability *p*<		
				Milk	Sample	Milk × Sample
Raw	Core	3.211 A	0.126	0.001	0.069	0.028
	Under-rind	3.147 A	0.126			
Pasteurized	Core	0.890 C	0.178			
	Under-rind	1.536 B	0.178			

On the column: A, B, C: $p \leq 0.01$.

Table 3 shows the simple statistics of ALP real values detected in Pecorino Siciliano cheese samples, which were produced by nine dairies that belong to the PDO Protection Consortium. In accordance with the classification proposed in Table 1, Table 3 shows two groups of results. Cheeses with a severe process showed a mean range between 26.5 and 909.1 mU/g, while cheeses cooked with a weak process showed a mean range between 1952.0 and 3148.6 mU/g. Even within the same cooking class, the variability between different dairies was wide. This is likely due to the different cooking systems practiced by different cheesemakers, who change the cooking system according to their daily needs or environmental temperatures. For this reason, a third group of cheeses was created (i.e., mixed). This group was produced by dairies that adapted a severe cooking system but functioned in different environmental conditions and used lower cooking temperatures. This cooking class presented a mean value similar to the weak category, i.e., equal to 1699.6 mU/g, but with a wide standard deviation, 1152.6 mU/g.

Table 3. ALP values in dairies belonging to the PDO Protection Consortium.

Dairy	Cooking Classes	Temperature of Cheese after Cooking (°C)	n.	Mean (mU/g)	SD (mU/g)	Max (mU/g)	Min (mU/g)
A	severe	54.5	5	26.5	14.9	47.7	13.0
B	severe	52.0	5	539.3	695.8	1756.1	11.3
C	severe	50.0	10	88.1	111.0	298.5	13.7
D	severe	47.0	4	839.0	1614.1	3714.8	25.2
E	severe	47.5	5	86.9	161.2	375.0	7.8
F	Severe	49.5	3	909.1	1198.8	2831.9	37.2
H	weak	44.7	12	2992.8	1533.8	5543.4	1433.3
I	weak	44.0	20	1952.0	1363.9	4150.2	512.6
L	weak	45.5	6	3148.6	382.7	3781.2	2682.4
Other PDO cheesemaker	mixed	45.0	8	1699.6	1152.6	3714.8	537.0
From the market	Pasteurized milk	40.0	20	12.3	5.9	23.9	5.8

Therefore, even when the same dairy is used during cheese production, different values of alkaline phosphatase can be recorded.

Pecorino Siciliano cheese samples that report on the lables "pasteurized milk product" present a mean value of 12.3 mU/g with a range between 5.8 and 23.9 mU/g. The ALP mean value of these cheeses is lower than Pecorino Siciliano PDO cheeses. This is true even for those produced under severe cooking conditions.

In this study, we reported the statistical analysis carried out on ALP logarithmic values (Table 4). The statistical model explains that 89% of variability (R^2) resulted in a statistically significant factor ($p < 0.001$). The logarithmic least square means (LSM) were statistically different between cooking classes. The LSM of cheese samples produced with pasteurized milk was 1.048, which corresponded to 15.7 mU/g. This was statistically lower than raw milk cheeses cooked under a weak (3.308; $p < 0.001$), severe (1.612; $p < 0.001$), or mixed cooking process (3.128; $p < 0.001$). For cheeses produced with raw milk, ALP values were classified as weak and were significantly higher than those in the severe cooking class. No differences were found with respect to the mixed class. ALP values in the severe cooking class were significantly lower than those in either the weak or mixed classes.

The above results appear to conflict with those reported in the literature [10]. However, few papers have analyzed the value of phosphatase on cheeses and, in particular, on traditional cheeses, because this method has only recently been validated [6,10,13].

Different cooking conditions determine the variability of alkaline phosphatase activity, such as the temperature of scotta whey before cooking, cooking time, the ratio between the liters of scotta whey used and the kg weight of the cheese. Severe cooking conditions for PDO Pecorino Siciliano cheeses determined low ALP values, which could potentially be misattributed to the thermal treatment of raw milk. In fact, in accordance with Egger et al. [10], these cheeses could have been produced by thermized milk. However, based on the control of cheese manufacturing procedures, the temperatures used in these dairies guarantees a more rigorous application of PDO protocol, ensuring that these values are sufficiently reliable.

Table 4. Least square means (LSM) of ALP logarithmic values.

Cooking Classes	ALP (log 10)	Standard Error	ALP (mU/g) *
Weak	3.308 A	0.055	2032.3
Severe	1.612 B	0.060	40.9
Mixed	3.128 A	0.120	1342.8
Pasteurized milk	1.048 C	0.076	15.7
Cooking classes	$p < 0.001$		$R^2 = 0.89$

On the column: A, B, C: $p \leq 0.01$; * these values were calculated as the inverse of logarithmic LSM.

A variability of ALP values was detected in Pecorino Siciliano cheese samples, particularly those produced with pasteurized milk. Several samples presented ALP values higher than 10 mU/g, a threshold reported by other authors [6,9,10]. These values discriminated against cheese produced with raw or pasteurized milk. A probable justification for this is that the production processes of these pecorino cheeses were not checked by us, and the sampling occurred on the basis of what was declared by the producer, who may have used different pasteurization temperatures.

4. Conclusions

In conclusion, we determined that the production of PDO Pecorino Siciliano cheese is characterized by a strong craftsmanship that indicates a wide variability between dairies. Different cooking systems that used scotta whey at various times and temperatures, and in different types of vats, showed a wide variability of ALP values and an unclear demarcation between Pecorino Siciliano cheeses produced with raw milk or pasteurized milk.

The presence of low ALP values detected in PDO Pecorino Siciliano samples that underwent severe cooking temperatures at the core of the cheese (i.e., above 47 °C), but were produced with raw milk, fall within the range of ALP values detected in pasteurized milk cheeses. The overlap of these values suggests that caution should be used when applying alkaline phosphatase as a control tool on raw or pasteurized milk cheeses. According to Clawin-Rädecker et al. [6], further scientific studies are necessary to establish whether there are threshold values capable of discriminating against the type of milk (raw or pasteurized) used for producing PDO Pecorino Siciliano.

Author Contributions: Conceptualization, M.T., V.C., and G.L.; methodology, V.L.P. and A.M.; software, M.T.; validation, M.T., V.C., and G.L.; formal analysis, M.T., V.C. and G.L.; investigation, V.L.P., A.M. and M.A.; resources, M.T.; data curation, V.L.P, A.M. and M.A.; writing—original draft preparation, M.T.; writing—review and editing, M.T. and V.C.; visualization, M.T. and V.C.; supervision, G.L.; project administration, M.T.; funding acquisition, M.T. All authors have read and agreed to the published version of the manuscript.

Funding: This work was supported by the AGER 2 Project, grant no. 2017-1144.

Institutional Review Board Statement: Not applicable.

Informed Consent Statement: Informed consent was obtained from all subjects involved in the study.

Data Availability Statement: The data presented in this study are available on request from the corresponding author.

Conflicts of Interest: The authors declare no conflict of interest.

References

1. Ricci, M. La Scoperta Degli Archeologi Siciliani: A Troina si Produceva Formaggio Oltre Anni fa (The Discovery of Sicilian Archaeologists: Cheese was Produced in Troina over Years ago). 2017. Available online: http://www.cronachedigusto.it/archiviodal-05042011/325-scenari/21156-la-scoperta-degli-archeologi-siciliani-a-troina-si-produceva-formaggio-oltre-5000-anni-fa.html (accessed on 10 April 2021).
2. Gaglio, R.; Todaro, M.; Settanni, L. Improvement of raw milk cheese hygiene through the selection of starter and non-starter lactic acid bacteria: The successful case of PDO pecorino siciliano cheese. *Int. J. Environ. Res. Public Health* **2021**, *18*, 1834. [CrossRef] [PubMed]
3. Rankin, S.A.; Christiansen, A.; Lee, W.; Banavara, D.S.; Lopez-Hernandez, A. Invited review: The application of alkaline phosphatase assays for the validation of milk product pasteurization. *J. Dairy Sci.* **2010**, *93*, 5538–5551. [CrossRef] [PubMed]
4. ISO. *ISO 11816-2: Milk and Milk Products E Determination OF Alkaline Phosphatase Activity—Part 2: Fluorometric Method for Cheese*; IOS Press: Amsterdam, The Netherlands, 2015.
5. Lechner, E.; Ostertag, S. Aktivität at der alkalischen Phosphat äse in milch und milchprodukten. Teil 2. *Dtsch. Milchwirtsch.* **1993**, *44*, 1146–1149.
6. Clawin-Rädecker, I.; De Block, J.; Egger, L.; Willis, C.; Felicio, M.T.D.S.; Messens, W. The use of alkaline phosphatase and possible alternative testing to verify pasteurisation of raw milk, colostrum, dairy and colostrum-based products. *EFSA J.* **2021**, *19*, e06576. [PubMed]
7. Bisig, W.; Fröhlich-Wyder, M.T.; Jakob, E.; Wechsler, D. Comparison between Emmentaler PDO and generic emmental cheese production in Europe. *Aust. J. Dairy Technol.* **2010**, *65*, 206.
8. Cattaneo, S.; Hogenboom, J.A.; Masotti, F.; Rosi, V.; Pellegrino, L.; Resmini, P. Grated Grana Padano cheese: New hints on how to control quality and recognize imitations. *Dairy Sci. Technol.* **2008**, *88*, 595–605. [CrossRef]
9. Pellegrino, L.; Tirelli, A.; Masotti, F.; Resmini, P. Significance of the main chemical indicators of heat load for characterizing raw, thermized, pasteurized and high temperature pasteurized milk [beta-lactoglobulin, alkaline phosphatase, lactoperoxidase]. In *Heat Treatments and Alternative Methods*; IDF Symposium: Vienna, Austria, 1996.
10. Egger, L.; Nicolas, M.; Pellegrino, L. Alkaline phosphatase activity in cheese as a tracer for cheese milk pasteurization. *LWT Food Sci. Technol.* **2016**, *65*, 963–968. [CrossRef]
11. Todaro, M.; Francesca, N.; Reale, S.; Moschetti, G.; Vitale, F.; Settanni, L. Effect of different salting technologies on the chemical and microbiological characteristics of PDO Pecorino Siciliano cheese. *Eur. Food Res. Technol.* **2011**, *233*, 931–940. [CrossRef]
12. SAS. SAS/STAT qualification tools user's guide (version 9.2). In *Statistical Analysis System*; Institute Inc.: Cary, NC, USA, 2010.
13. Yoshitomi, K. Alkaline phosphatase activity in cheeses measured by fluorometry. *Int. J. Food Sci. Technol.* **2004**, *39*, 349–353. [CrossRef]

Article

Optimum DNA Extraction Methods for Edible Bird's Nest Identification Using Simple Additive Weighting Technique

Meei Chien Quek [1], Nyuk Ling Chin [1,*] and Sheau Wei Tan [2]

[1] Department of Process and Food Engineering, Faculty of Engineering, Universiti Putra Malaysia (UPM), Serdang 43400, Malaysia; qian_mc@hotmail.com

[2] Laboratory of Vaccines and Immunotherapeutics, Institute of Bioscience, Universiti Putra Malaysia (UPM), Serdang 43400, Malaysia; tansheau@upm.edu.my

* Correspondence: chinnl@upm.edu.my; Tel.: +60-39-769-6353

Citation: Quek, M.C.; Chin, N.L.; Tan, S.W. Optimum DNA Extraction Methods for Edible Bird's Nest Identification Using Simple Additive Weighting Technique. *Foods* **2021**, *10*, 1086. https://doi.org/10.3390/foods10051086

Academic Editor: Raúl González-Domínguez

Received: 8 April 2021
Accepted: 4 May 2021
Published: 14 May 2021

Publisher's Note: MDPI stays neutral with regard to jurisdictional claims in published maps and institutional affiliations.

Copyright: © 2021 by the authors. Licensee MDPI, Basel, Switzerland. This article is an open access article distributed under the terms and conditions of the Creative Commons Attribution (CC BY) license (https://creativecommons.org/licenses/by/4.0/).

Abstract: A simple additive weighting (SAW) technique was used to determine and compare the overall performance of five DNA extraction methods from conventional (SDS method) to commercial kits (Qiagen, Wizard, and NucleoSpin) for identifying origins of edible bird's nest (EBN) using end-point polymerase chain reaction (PCR). A hybrid method (SDS/Qiagen) which has been developed by combining the conventional SDS method with commercialised Qiagen was determined as the most suitable in terms of speed and cost-effectiveness. The determination of optimum extraction method was by the performances on efficiency and feasibility, extracted DNA concentration, purity, PCR amplifiability, handling time and safety of reagents used. The hybrid SDS/Qiagen method is less costly compared to the commercial kits and offered a more rapid alternative to the conventional SDS method with significant improvement in the yield, purity and PCR amplifiability. The developed hybrid SDS/Qiagen method provides a more practical alternative over the lengthy process using conventional method and expensive process using commercial kits. Using the simple additive weighting (SAW) technique and analysis, the Qiagen method is considered the most efficient and feasible method without consideration of cost as it yielded the purest extracted DNA and achieved the highest PCR amplifiability with the shortest turnaround time.

Keywords: SDS method; Qiagen method; polymerase chain reaction (PCR); multiple attribute decision making (MADM) analysis; *Aerodramus*

1. Introduction

Edible bird's nest (EBN), also known as cubilose, is one of the most precious and expensive food produced from saliva of two swiftlet species, *Aerodramus fuciphagus* and *Aerodramus maximus* [1]. It is mainly originated from Southeast Asia countries, such as Indonesia, Thailand, Malaysia, and Vietnam [2]. Due to high demand and high price of genuine EBN, counterfeit and adulterated EBN are increasingly rampant in the markets. This has raised awareness of the importance of authentication of EBN. Several studies have employed DNA-based method to identify genuineness of EBN and its products [1,3,4]. The DNA-based method is known to be the most appropriate tool to identify species present in food [1] because DNA strands serve as templates for building new copies in cell replication, repair, and transcription. The DNA-based method is relatively faster, has greater sensitivity and specificity compared to the analytical and chemical methods when it comes to retrieving genetic information from food materials for species identification, varieties discrimination and allergy diagnosis [5–7].

Polymerase chain reaction (PCR) is a commonly used DNA-based method for identification and detection of food adulterant [8]. The mitochondrial cytochrome *b*, 12S rRNA and 16S rRNA genes are most widely used genetic markers for species identification by PCR due to availability of reference sequences in databases [9]. PCR method was developed to identify plant and insect origins of bee honey where markers of mitochondrial, nuclear, and chloroplast DNA were used to differentiate honey based on its origin [10]. A wide variety of meat products from different species like cattle, buffalo, sheep, goat, pig, chicken, ostrich, turkey, and rabbit were also authenticated by sequencing PCR products from a 555 bp region of mitochondrial cytochrome b gene [11]. Out of 20 commercial fresh and precooked products, 15% of them were found to be mislabeled. This method has also been applied for species identification of dairy products [12], fish [13] and meat [14], and detection of fruit ingredients in juices [15]. Despite being an accurate and efficient identification method, DNA-based method often faced challenges in terms of quality and quantity of extracted DNA which rely heavily on DNA extraction method. An efficient and reliable DNA extraction method must be effective in yielding adequate amount of high-quality extracted DNA and suitable for subsequent downstream molecular analyses such as conventional/end point PCR, real-time PCR, and DNA microarrays [16–18]. Various studies that have evaluated and compared DNA extraction methods on different subject matters are available [19–22].

As EBN naturally contains low amount of DNA, it is extremely challenging to extract good quality and sufficient quantity of DNA from EBN. The presence of abundant glycoprotein increases the difficulty to obtain high quality DNA [4]. The use of commercial kits are expensive while conventional methods are tedious, lengthy and hazardous. Lin et al. [4] employed two conventional methods, i.e., modified sodium dodecyl sulphate (SDS) and cetyltrimethylammonium bromide (CTAB) methods to overcome the two challenges in extracting DNA from EBN. Although the modified method can deliver good results, it was very time consuming and involved using sodium dodecyl sulphate reagent that can cause great hazard to human health. While existing DNA extraction protocols are available, they have not been compared comprehensively, specifically on EBN.

This work aimed to compare and determine the best method for efficient and feasible DNA extraction method for rapid species identification of EBN using a systematic analysis and engineering approach known as the simple additive weighting (SAW) technique. It is classified as a multiple attribute decision making (MADM) analysis. The hybrid SDS/Qiagen method, which is new, rapid, and cost-effective alternative was evaluated and compared with SDS method and three commercially available kits including Wizard Magnetic DNA purification system for food kit, NucleoSpin food kit, and DNeasy mericon food kit in terms of extracted DNA concentration, purity and PCR amplifiability, plus the time, cost, and safety of the extraction method. The optimal DNA extraction method for EBN was identified using simple additive weighting technique and validated for applicability for species identification of EBN through end-point PCR.

2. Materials and Methods

2.1. Edible Bird's Nest Preparation

The 13 types of EBN samples originated from two swiftlet species, *A. fuciphagus* and *A. maximus* were collected from Malaysia (Table 1). The 11 unprocessed EBN samples were obtained directly from local farmers and two processed EBN samples were purchased from local markets. Processed EBNs have undergone harvesting, sorting, soaking, cleaning, moulding, drying, and packaging processes. The unprocessed EBNs were cleaned manually using tweezers to remove loose feathers and impurities. The EBN samples were then pulverised with liquid nitrogen using mortar and pestle, sieved through 1 mm mesh size to obtain a homogenous and fine powder for optimum yield. The samples were stored at 4 °C until DNA extraction. Two fake EBN samples were also used as samples and they were subsequently omitted in analysis due to negative results of extracted DNA.

2.2. DNA Extraction

Total genomic DNA of EBN samples were extracted using five different DNA extraction methods, namely Wizard (Promega Corporation, Madison, WI, USA), NucleoSpin (Macherey-Nagel GmbH and Co. KG, Düren, Germany), Qiagen (Qiagen Corporation, Hilden, Germany), SDS, and SDS/Qiagen. Each EBN was extracted in quadruplicate to ensure reproducibility of the extraction methods. For fair comparison of all extraction methods, the amount of starting materials was standardised to 25 mg of EBN samples and the final volume of extracts was fixed at 100 µL. The extracted DNA was stored at −20 °C.

2.3. Wizard Magnetic DNA Purification System for Food Kit (Wizard Method)

The EBN samples were extracted using commercial kit, Wizard® Magnetic DNA purification system for food (Promega Corporation, Madison, WI, USA) following the manufacturer's instructions except the volume adjustments in lysis buffers. Each EBN of 25 mg was vigorously vortexed with 450 µL of Lysis Buffer A and 5 µL of RNase A, then vortexed again with 200 µL of Lysis Buffer B for 15 s in a 1.5 mL microcentrifuge tube. The tube was laid on its side and incubated at room temperature for 10 min. The sample was vigorously vortexed with 700 µL of precipitation solution and centrifuged at $13,000 \times g$ for 10 min in a 5415D microcentrifuge (Eppendorf, Hamburg, Germany) for protein precipitation. About 700 µL of supernatant was vortexed with 50 µL of resuspended MagneSil™ paramagnetic particles (PMP) in a new microcentrifuge tube, and then continued with remaining procedures in the manufacturer's instructions. The Wizard method lyses with guanidine thiocyanate and RNase, and binds DNA to silica-coated magnetic beads.

2.4. NucleoSpin Food Kit (NucleoSpin Method)

NucleoSpin® food kit (Macherey-Nagel GmbH and Co. KG, Düren, Germany) was performed with minor modifications by doubling the lysis buffers volume and prolonging the incubation time. About 25 mg of each EBN was mixed with 1100 µL of preheated Lysis Buffer CF at 65 °C and 20 µL of proteinase K. The sample was incubated at 65 °C for 1 h in a ThermoStat plus heating block (Eppendorf, Hamburg, Germany) and centrifuged at $11,000 \times g$ for 10 min to pellet contaminations and cell debris. Then, 450 µL of clear supernatant was vortexed with 450 µL of binding buffer C4 and 450 µL of 96% (v/v) ethanol, and followed with DNA binding, washing, and elution steps in the manufacturer's instructions. The NucleoSpin method lyses with chaotropic salts, denaturants, detergents, and proteinase K, and binds DNA to silica membrane in spin column.

2.5. DNeasy Mericon Food Kit (Qiagen Method)

The Qiagen method was conducted using DNeasy® mericon™ food kit (Qiagen GmbH, Hilden, Germany) following the manufacturer's instructions with slight alterations. Each EBN of 25 mg was vortexed with increased volume of 1.3 mL food lysis buffer and 5 µL proteinase K, and then incubated for a longer period of 1 h at 60 °C to enhance inhibitor precipitation. The following extraction procedures were proceeded with the manufacturer's instructions until the elution step, where DNA was eluted from QIAquick spin column with 100 µL of buffer EB instead of 150 µL for standardisation purpose. The Qiagen method lyses with non-ionic detergent CTAB and proteinase K, and binds DNA to silica membrane in spin column.

2.6. Conventional SDS Method (SDS Method)

SDS method was performed following Lin et al. [23] with some modifications. About 25 mg of each EBN was added with 1.2 mL of lysis buffer (10 g/L SDS, 50 mM Tris-HCl pH 8.0, 10 mM EDTA pH 8.0, 0.04 M DTT, 200 mg/L proteinase K, 2.0 M NaCl preheated at 65 °C) in a microcentrifuge tube. The mixture was vortexed and incubated at 65 °C for 1 h, followed by centrifugation at $12,000 \times g$ for 5 min at 4 °C to remove undigested debris. A 1000 µL of supernatant was transferred to a new tube containing equal volume of chloroform/isoamyl alcohol solution (24:1) and mixed well before centrifuged to remove protein. Supernatant of 500 µL was added with 50 µL of 10% CTAB/0.7 M NaCl buffer preheated at 65 °C and incubated at room temperature for 15 min, then mixed with 500 µL chloroform/isoamyl alcohol solution. The mixture was centrifuged to remove remaining CTAB and glycoprotein, and 400 µL of supernatant was transferred to new tube. The supernatant was mixed with 280 µL of cold isopropanol and centrifuged for DNA precipitation. The pellet was washed with 1 mL of 75% (v/v) ethanol and centrifuged at 12,000× g for 5 min. The DNA pellet was air dried and resuspended in 100 µL of nuclease-free water. The SDS method lyses with anionic detergent SDS, DTT and proteinase K, purifies DNA with cationic detergent CTAB and chloroform/isoamyl alcohol, and precipitates DNA with cold isopropanol.

2.7. Hybrid SDS Method and Qiagen Method (SDS/Qiagen Method)

Hybrid SDS/Qiagen method was developed by combining the SDS method and Qiagen method. The EBN samples were lysed using the SDS method, and then the following steps of DNA binding, washing, and elution were performed using QIAquick spin column from the Qiagen method. The initial procedure of this method was similar to the SDS method, from sample lysis to the addition of cold isopropanol steps. After mixing 400 µL of supernatant with 280 µL of cold isopropanol, the mixture was transferred to the spin column and then proceeded with remaining procedures in the Qiagen method. This SDS/Qiagen method lyses with anionic detergent SDS, DTT, and proteinase K, purifies DNA with cationic detergent CTAB and chloroform/isoamyl alcohol, and binds DNA to silica membrane in spin column.

2.8. DNA Quantification and Purity

DNA concentration of EBN samples was quantified with spectrophotometric assay by measuring UV absorbance at 260 nm (A_{260}) using a BioSpectrometer® kinetic spectrophotometer (Eppendorf, Hamburg, Germany). One-way analysis of variance (ANOVA) was performed to compare the DNA concentration between five different DNA extraction methods (Table 1). Significant differences between means were evaluated using Tukey's test at a confidence level of 95%. Fluorometric quantification assay was also performed based on fluorescent DNA binding dyes using a Qubit® 2.0 fluorometer (Life Technologies, Carlsbad, CA, USA) and Qubit® dsDNA high sensitivity assay kit. This assay is highly specific and selective for double-stranded DNA (dsDNA) quantification. Purity of the extracted DNA was determined by the absorbance ratios of 260 and 280 nm (A_{260}/A_{280}) using the spectrophotometer.

2.9. PCR Amplification

PCR amplification was performed to compare the performance of five different DNA extraction methods. The extracted DNA of EBN samples were amplified using mitochondrial cytochrome b gene primers available in the literature, L15302 (5′ GTA GGA TAT GTC CTN CCH TGA GG 3′) and H15709 (5′ GGC ATA TGC GAA TAR GAA RTA TCA 3′) to amplify 406 bp PCR products (S1) [24]. These primers were synthesised by AIT Biotech in Singapore. PCR amplification was conducted in a 50 µL total reaction volume containing final concentration of 1 x MyTaq™ Mix PCR buffer (Bioline, London, UK), 0.4 µM of each forward and reverse primer, and 0.02–1.70 ng/µL of DNA template. A mixture with no DNA template was used as negative control. The amplification was performed using a C1000 Touch™ thermal cycler (Bio-Rad, Hercules, CA, USA) with the following PCR cycle: initial denaturation at 95 °C for 3 min; followed by 35 cycles of denaturation at 95 °C for 15 s, primers annealing at 53 °C for 30 s and extension at 72 °C for 30 s; then final extension was conducted at 72 °C for 5 min. Each PCR amplification was performed in at least triplicate to ensure its repeatability.

PCR products were analysed by agarose gel electrophoresis using a 1.5% (w/v) agarose gel pre-stained with Red-Safe™ DNA dyes (iNtRON Biotechnology, Sungnam, Korea) at 80 V for 60 min. A 100 bp HyperLadder™ DNA ladder (Bioline, London, UK) was used as PCR products size marker. The gel was visualised under UV light using a Gel™ Doc XR imaging system (Bio-Rad, Hercules, CA, USA). The expected PCR product size was 406 bp.

2.10. DNA Sequencing

The PCR products were sequenced using an ABI3730x1 automated DNA sequencer (Applied Biosystems, Foster City, CA, USA) and the same primers used in PCR amplification. The nucleotide sequences obtained were subjected to the nucleotide basic local alignment search tool (BLASTN) available at National Centre for Biotechnology Information (NCBI) (http://blast.ncbi.nlm.nih.gov/Blast.cgi (accessed on 25 May 2015)) for sequence similarity search. Generally, if PCR product sequence and database sequences show maximum identities or highest similarities, the identity of EBN samples can be confirmed.

2.11. Ranking of DNA Extraction Method Using Simple Additive Weighting Technique

Multiple attributes including dsDNA concentration, purity, PCR amplifiability, procedure simplicity, safety of reagents (beneficial attributes), and handling time (non-beneficial attribute) were used to evaluate overall performance of DNA extraction method. Cost was evaluated based on a relative qualitative scale. As some of the attributes are contradictory, it increased the difficulty in selecting the optimal extraction method. A simple additive weighting (SAW) technique of multiple attribute decision making (MADM) analysis was used. This technique clustered all attributes results into a comprehensive system based on mathematical scoring technique thus provided ranking to each DNA extraction method [25]. All attributes used were normalised to standardised values to ensure they contribute evenly to a scale for comparison purposes [26]. The level of procedure simplicity and reagents safety attributes were rated based on direct rating method [27], where 1 and 2 indicated simple and more simple for procedure simplicity, and safe and more safe for reagents safety. Each attribute was given a weightage as an indication of its importance in DNA extraction method selection and the sum of all weightage is equal to 1. The SAW technique was performed using the following equation [28]:

$$S_i = \sum_{j=1}^{6} w_j r_{ij} \ for \ i = 1, 2, 3, 4, 5$$

where S_i is overall score, w_j is weightage of jth attribute, and r_{ij} is standardised value of the ith DNA extraction method with respect to the jth attribute. For beneficial attribute, $r_{ij} = x_{ij}/x_{ij(\max)}$ and for non-beneficial attribute, $r_{ij} = x_{ij(\max)}/x_{ij}$, where x_{ij} is original value and $x_{ij(\max)}$ is the largest value of the jth attribute of the ith DNA extraction method [29,30]. DNA extraction method with the highest overall score was granted the highest ranking, hence identified as the optimal extraction method.

3. Results

3.1. DNA Concentration

Table 1 shows DNA concentration of 13 EBN samples extracted using five different DNA extraction methods and measured using spectrophotometric assay. Regardless of extraction methods used, the processed EBNs (samples 12–13) which have undergone intensive degree of processing generally have shown lower DNA concentration than the unprocessed EBNs samples (samples 1–11). Among the three commercial kits tested, the Wizard and Qiagen methods gave significantly highest DNA concentration ($p < 0.05$) of 2.23–5.03 ng/µL and 1.35–4.50 ng/µL, respectively. Interestingly, in Qiagen method, four EBNs (samples 8–11) which originated from *A. maximus* produced significantly lower amount of extracted DNA than others ($p < 0.05$). NucleoSpin method yielded significantly lowest DNA concentration for EBN samples ($p < 0.05$) ranging from 0.30 to 1.25 ng/µL. The SDS method, which is a standard method and widely used in DNA extraction of EBN [4,31] however, gave relatively low DNA concentration in this study. Despite its low amount of extracted DNA, the SDS method showed significantly greater ability in extracting DNA from *A. maximus* EBNs (samples 8–11) than *A. fuciphagus* EBNs ($p < 0.05$). The hybrid SDS/Qiagen method showed a significant improvement in DNA recovery compared to the SDS method with at least 2-fold's increment. It yielded significantly highest DNA concentration for EBN samples ($p < 0.05$) ranging from 4.18 to 5.68 ng/μL.

Table 1. DNA concentration of EBN samples extracted with five DNA extraction methods as measured by spectrophotometry.

EBN	Description			DNA Concentration (ng/μL) †				
	Type	Species ‡	Origin	Wizard	NucleoSpin	Qiagen	SDS	SDS/Qiagen
1	Unprocessed	A. fuciphagus	* Segamat, Johor	3.65 ± 0.21 [b]	1.25 ± 0.35 [c]	4.33 ± 0.33 [ab]	1.10 ± 0.18 [c]	4.60 ± 0.42 [a]
2	Unprocessed	A. fuciphagus	* Kapar, Selangor	4.40 ± 0.54 [a]	0.83 ± 0.05 [b]	3.65 ± 0.61 [a]	1.58 ± 0.15 [b]	4.23 ± 0.72 [a]
3	Unprocessed	A. fuciphagus	* Nibong Tebal, Penang	3.33 ± 0.13 [ab]	0.45 ± 0.13 [c]	3.23 ± 0.90 [b]	1.35 ± 0.06 [c]	4.43 ± 0.73 [a]
4	Unprocessed	A. fuciphagus	* Klang, Selangor	3.55 ± 0.13 [c]	0.60 ± 0.08 [e]	4.08 ± 0.25 [b]	1.35 ± 0.13 [d]	4.78 ± 0.21 [a]
5	Unprocessed	A. fuciphagus	** Sarikei, Sarawak	3.80 ± 0.81 [a]	0.60 ± 0.08 [b]	4.50 ± 0.26 [a]	1.10 ± 0.08 [b]	4.45 ± 0.19 [a]
6	Unprocessed	A. fuciphagus	** Gomantong Cave, Sabah	2.23 ± 0.43 [c]	0.50 ± 0.08 [d]	3.75 ± 0.53 [b]	0.70 ± 0.00 [d]	4.98 ± 0.44 [a]
7	Unprocessed	A. fuciphagus	** Baram, Sarawak	5.03 ± 0.78 [a]	1.23 ± 0.15 [c]	3.38 ± 0.10 [b]	1.83 ± 0.10 [c]	4.65 ± 0.70 [a]
8	Unprocessed	A. maximus	** Gomantong Cave, Sabah	3.20 ± 0.87 [b]	0.45 ± 0.13 [d]	1.35 ± 0.26 [cd]	1.65 ± 0.19 [c]	5.18 ± 0.68 [a]
9	Unprocessed	A. maximus	** Niah Cave, Sarawak	3.30 ± 0.70 [b]	0.68 ± 0.10 [d]	1.55 ± 0.13 [cd]	1.73 ± 0.15 [c]	5.10 ± 0.75 [a]
10	Unprocessed	A. maximus	** Niah Cave, Sarawak	4.30 ± 0.61 [a]	0.35 ± 0.06 [c]	1.90 ± 0.46 [b]	2.28 ± 0.56 [b]	5.68 ± 1.08 [a]
11	Unprocessed	A. maximus	** Subis Cave, Sarawak	4.50 ± 0.60 [a]	0.55 ± 0.06 [c]	1.58 ± 0.22 [b]	2.15 ± 0.21 [b]	4.75 ± 0.49 [a]
12	Processed	A. fuciphagus		2.55 ± 0.37 [b]	0.53 ± 0.10 [c]	4.28 ± 0.33 [a]	0.73 ± 0.15 [c]	4.18 ± 0.50 [a]
13	Processed	A. fuciphagus		2.95 ± 0.13 [b]	0.30 ± 0.00 [d]	1.53 ± 0.13 [c]	1.40 ± 0.14 [c]	4.50 ± 0.16 [a]
			Average	3.59 ± 0.49	0.64 ± 0.11	3.01 ± 0.35	1.46 ± 0.16	4.73 ± 0.54

‡ A. fuciphagus, Aerodramus fuciphagus; A. maximus, Aerodramus maximus. † Values are mean ± standard deviation with $n = 4$ and different superscript letters in the same row indicate significantly different ($p < 0.05$). * Peninsular Malaysia; ** East Malaysia.

Comparing quantification assays, the average DNA and dsDNA concentrations of all EBN samples from five different DNA extraction methods quantified by spectrophotometry and fluorometry, respectively, are shown in Figure 1. The average DNA concentration via SDS/Qiagen method was significantly highest ($p < 0.05$) at 4.73 ng/μL by absolute value, followed by Wizard, Qiagen, SDS and NucleoSpin methods using the spectrophotometric quantification. For fluorometric quantification, the SDS method gave the highest dsDNA concentration while the Wizard method yielded the lowest for all EBN samples.

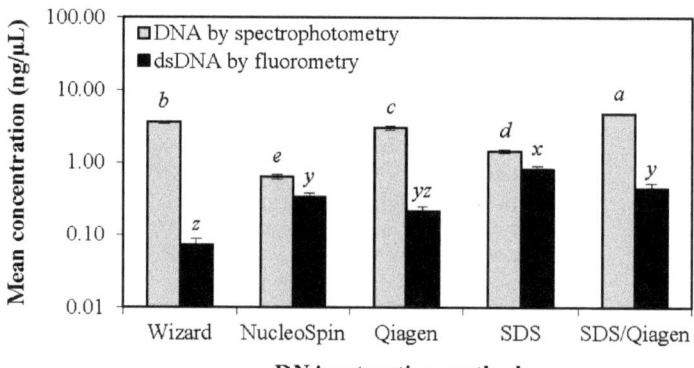

Figure 1. Mean DNA and dsDNA concentrations of EBN samples extracted with five different DNA extraction methods as measured by spectrophotometry and fluorometry, respectively. Different letters in each quantification method indicate significant differences ($p < 0.05$). Values are mean ± standard error with samples size $n = 52$ (spectrophotometry) and $n = 13$ (fluorometry).

3.2. DNA Purity

Figure 2 shows the purity of extracted DNA from EBN samples determined by the absorbance ratio of 260 and 280 nm (A_{260}/A_{280}) where A_{260} and A_{280} values indicated the presence of DNA and protein, respectively. The extracted DNA is considered pure if A_{260}/A_{280} value ranged between 1.7 and 2.0 [32]. The Wizard and Qiagen methods obtained the highest DNA purity. A closer observation showed that Qiagen method had a higher sampling fraction of 6/13 than the Wizard method with 4 samples out of 13. Contrarily, the NucleoSpin, SDS and SDS/Qiagen methods gave relatively low DNA purity ranging from 0.87 to 1.42. Figure 2 also shows that the purity of extracted DNA was not significantly different between processed and unprocessed EBNs in any extraction method suggesting processing of EBN does not affect DNA purity.

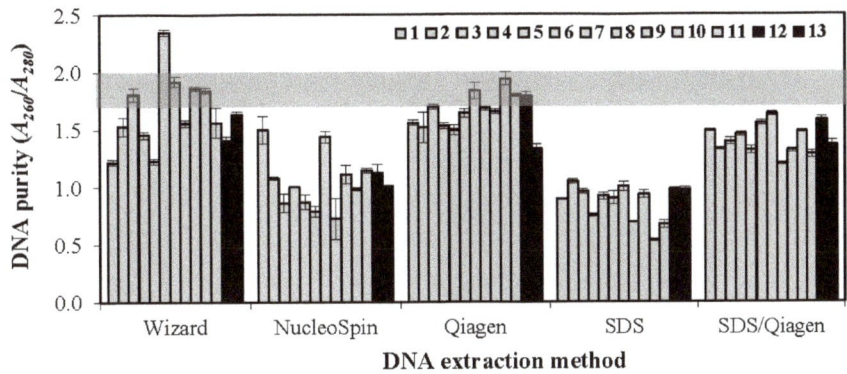

Figure 2. Comparison of purity of DNA extracted from 13 EBN samples with five different DNA extraction methods. Samples 1–11 are unprocessed EBNs and samples 12–13 are processed EBNs. Grey shaded area represents satisfactory range for pure DNA from 1.7 to 2.0. Values are mean ± standard error with samples size.

3.3. PCR Amplifiability

Figure 3 shows PCR amplification results using a pair of cytochrome b gene primers at expected size of 406 bp. The extracted DNA of unprocessed EBNs was successfully amplified while the processed ones (lanes 12–13) showed relatively faint PCR bands. The Wizard method gave no visible lanes 12 and 13. Weak PCR bands appeared in the NucleoSpin (lane 13) and SDS (lane 12) while Qiagen and SDS/Qiagen gave reasonable PCR bands for lanes 12 and 13. From the five DNA extraction methods, only DNA extracted with Qiagen method gave consistently intense PCR bands with expected size for all EBN samples.

3.4. Time, Safety, and Economic Evaluation of Extraction Methods

Based on a single sample handling [33], commercial kits required less time for DNA extraction than the SDS and SDS/Qiagen methods (Table 2). From the five DNA extraction methods, the commercial kits employed less hazardous reagents than SDS and SDS/Qiagen methods which required the use of corrosive and flammable reagents such as SDS, CTAB, chloroform/isoamyl alcohol and isopropanol. Most of the reagents used in all five DNA extraction methods were classified as skin and eyes irritant, and they were less likely to cause harmful effects if handled with care [34]. The most economical DNA extraction method was the SDS method, followed by SDS/Qiagen method. The reagents used in SDS method were common and often purchased in bulk quantity, thus it is cheapest in extraction cost. The three commercial kits were the most expensive. Comparing between the commercial kits, the Qiagen method had the lowest extraction cost with estimation of USD 3.00 for one sample, followed by Wizard and NucleoSpin methods at USD 3.40 and

USD 4.00, respectively. The cost for a single sample DNA extraction has been estimated based on average reagents and commercial kits prices in Malaysia. The cost attribute was not included in this DNA extraction method selection due to subjectivity of reagent costs for the SDS and SDS/Qiagen methods.

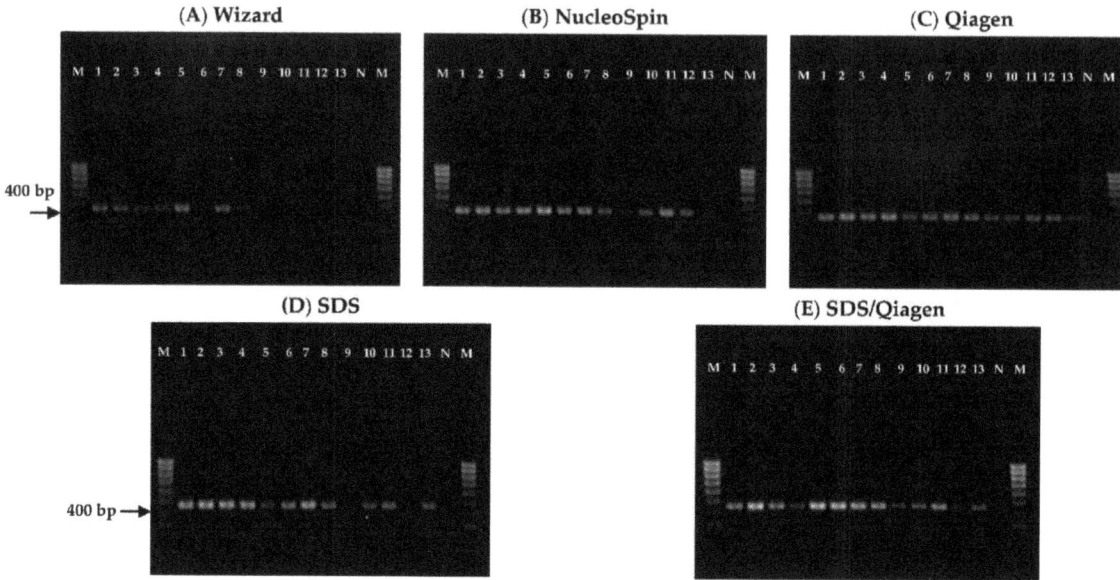

Figure 3. Gel electrophoreses of the 406 bp PCR products of cytochrome b gene amplified from extracted DNA of EBN samples with five different DNA extraction methods, namely (**A**) Wizard method, (**B**) NucleoSpin method, (**C**) Qiagen method, (**D**) SDS method and (**E**) hybrid SDS/Qiagen method. Lane M, 100 bp DNA ladder; lane 1–11, unprocessed EBNs; lane 12–13, processed EBNs; lane N, no template control.

Table 2. Evaluation of five different DNA extraction methods using simple additive weighting technique.

DNA Extraction Method	Attribute/Measured Data						Overall Score	Rank
	dsDNA ‡	Purity †	PCR †	Time ‡	Simplicity Φ	Safety Φ		
Weightage (Σ = 1)	1/6	1/6	2/6	1/6	1/6	1/6		
Wizard	0.08	0.31	0.54	2.0	2	2	1.02	3
NucleoSpin	0.34	0.00	0.85	2.0	2	2	1.06	2
Qiagen	0.22	0.46	1.00	2.0	2	2	1.25	1
SDS	0.81	0.00	0.85	4.5	1	1	0.78	4
SDS/Qiagen	0.44	0.00	0.92	4.0	1	1	0.75	5

‡ dsDNA concentration measured by fluorometry (ng/mL); handling time (hours). † Sampling fraction with purity between 1.7 and 2.0 or successful PCR amplification with intense bands, respectively. Φ Direct rating of procedure simplicity, 1, simple; 2, more simple, and reagents safety, 1, safe; 2, more safe.

3.5. Optimal DNA Extraction Method with SAW Technique

Table 2 shows that Qiagen method ranked first, followed by NucleoSpin, Wizard, SDS and SDS/Qiagen methods. The Qiagen method was identified as the most efficient and feasible DNA extraction method for EBN, yielding the highest success rate of PCR amplification with intense bands and excellent DNA purity, highest procedure simplicity and reagents safety, and required least handling time for DNA extraction. The most widely used conventional and standard method, the SDS was ranked the fourth. Despite obtaining highest amount of DNA, the SDS method gave the lowest DNA purity with relatively lengthy and tedious extraction procedure, and it also involved hazardous reagents.

3.6. Validation of Optimised Qiagen Method for Species Identification of EBN

The Qiagen method was validated to ensure its extracted DNA has the quality for downstream molecular applications. Table 3 shows that all 13 PCR products of the 406 bp cytochrome b gene sequences from EBN samples that were sequenced and subjected to BLASTN homology search were 100% identical to their respective published swiftlet sequences obtained from GenBank database. All the sequences of EBN samples were aligned to their respective swiftlet species sequences, *A. fuicphagus* or *A. maximus* available in GenBank database. These matching obtained BLASTN hits of 100% identity and E-values (Expected values) of 0 indicating that the hits were significantly matched.

Table 3. BLAST results on GenBank with first hit sequence using 406 bp of cytochrome b gene marker.

EBN	First Hit Sequence (Species and Accession Number)	Maximum Identity (%)	E-Value [†]
1	*Aerodramus fuciphagus* (JQ353840.1)	100%	1×10^{-87}
2	*Aerodramus fuciphagus* (JQ353840.1)	100%	1×10^{-87}
3	*Aerodramus fuciphagus* (JQ353840.1)	100%	1×10^{-87}
4	*Aerodramus fuciphagus* (JQ353840.1)	100%	1×10^{-87}
5	*Aerodramus fuciphagus* (JQ353840.1)	100%	1×10^{-87}
6	*Aerodramus fuciphagus* (JQ353840.1)	100%	1×10^{-87}
7	*Aerodramus fuciphagus* (JQ353840.1)	100%	1×10^{-87}
8	*Aerodramus maximus* (JQ353847.1)	100%	0.0
9	*Aerodramus maximus* (JQ353847.1)	100%	0.0
10	*Aerodramus maximus* (JQ353847.1)	100%	0.0
11	*Aerodramus maximus* (JQ353847.1)	100%	0.0
12	*Aerodramus fuciphagus* (JQ353840.1)	100%	1×10^{-87}
13	*Aerodramus fuciphagus* (JQ353840.1)	100%	1×10^{-87}

[†] E-value the number of hits one can "expect" to see by chance when searching a database of a particular size on BLAST search.

4. Discussion

The lower DNA concentration of processed EBNs may be related to DNA deterioration during processing which typically involves overnight soaking and drying of EBN [4]. It was evident that thermal processing of drying, cooking, baking, and roasting can cause DNA degradation in foods [23]. This trend was consistent with the findings published by Pirondini et al. [33] and Besbes et al. [35], who have reported higher amount of DNA in fresh milk and seafood than in their processed products. The DNA concentration was consistently higher than the dsDNA concentration, regardless of samples and extraction methods used (Figure 1). This could be due to the overestimation of DNA concentration by spectrophotometry as UV absorbance measurement are not selective and cannot distinguish DNA, RNA, or protein [36,37]. The fluorometric assay is also known to be more sensitive and specific for dsDNA only via fluorescent dyes binding, and it minimises the interference of RNA, protein and aromatic compounds in the extracted DNA [38]. As the fluorometric quantification provided a more selective, sensitive, and accurate method for quantifying nucleic acids than the spectrophotometric quantification, the dsDNA concentration was selection for subsequent extraction process.

In terms of DNA purity, Qiagen method was more superior in removing protein contaminants and inhibitors from EBN when compared with NucleoSpin, SDS, and SDS/Qiagen. This could probably due to protein contamination and organic solvents carryover in the extracted DNA of EBN samples. Generally, protein contamination and residual reagents such as ethanol, phenol, and chloroform interfere the A_{260}/A_{280} values and reduce the purity values to below 1.7 [34,39]. The residual reagents contamination may be effectively removed while maintaining the assay sensitivity using commercial nucleic acid extraction kit reagents such as GenElute Maxiprep binding columns [40]. The DNA purity from NucleoSpin and SDS methods may be optimised by adding filtration step with QIAquick spin column from Qiagen kit.

In using SAW technique to select the optimal DNA extraction method for EBN, the multiple contradictory attributes comprising dsDNA concentration, purity, PCR amplifiability, handling time, procedure simplicity, and reagents safety, PCR amplifiability was assigned with a higher weightage than other attributes because a successful PCR amplification is crucial for the subsequent molecular analysis, such as DNA sequencing [41]. It was necessary to consider other attributes of optimum DNA extraction method, such as handling time, procedure simplicity, safety of reagents used [5,42], and costs besides extracted DNA quality and quantity. The expensive cost of commercial kits is related to its sophisticated reagents and columns that are covered by international patents [33]. The handling time is directly proportional to procedure simplicity, where commercial kits contained simply fewer steps in extraction procedure than the SDS and SDS/Qiagen methods have shorter handling time. Most of the reagents needed were readily provided in the commercial kits. DNA extraction techniques employed in commercial kits was simpler than precipitation technique used in conventional SDS method, i.e., silica paramagnetic particles-based technique in Wizard method and column-based technique in NucleoSpin and Qiagen methods. Safety of reagents was evaluated following the material and safety datasheet (MSDS). In brief, commercial kits were fast, simple and safe but expensive whereas conventional methods were slow, tedious, and hazardous, but economical. Hybrid method was safer and faster than conventional methods, and less expensive than commercial kits.

The Qiagen method, found to be the optimal extraction method for EBN in this study, however, contradicts Wu et al. [1] who reported that NucleoSpin method was their best in EBN studies for successful PCR amplification. The difference in findings may be due to the variation in sample, DNA extraction method or the targeted gene of interest used for amplification. This is shown in this study when different extraction methods were suitable for DNA extraction of different species of EBN. The Qiagen method was more suitable for *A. fuciphagus* than *A. maximus* whereas the SDS method showed significantly greater ability in extracting EBN's DNA from *A. maximus* than *A. fuciphagus*. This may be due to the different nature and composition of food from different species which affected the DNA extraction [34]. The best PCR amplifiability results by Qiagen method may be attributed to the higher quality of DNA extracted. The weak PCR bands could be due to DNA degradation and fragmentation which occurred during EBN processing [4]. Nonetheless, the success rate of PCR amplification was not correlated to the concentration and purity of extracted DNA of EBN. For instance, the Wizard method yielded high amount of DNA with good purity but it gave relatively faint PCR bands for most of the amplified EBN samples. This observation is in agreement with previous work by Turci et al. [38], who reported unsuccessful amplification in most of the tomatoes products although high amounts of extracted DNA was yielded.

5. Conclusions

The SAW analysis has helped in determining optimal DNA extraction method for EBN species identification through end-point PCR. The hybrid DNA extraction method (SDS/Qiagen) was developed by replacing the DNA precipitation step with QIAquick spin column from the Qiagen method to improve DNA recovery of the SDS method which has shown great improvement as the silica-based column has greater DNA binding ability in the presence of chaotropic salts more efficiently. The hybrid method provides an alternative for a lower cost method than the commercial kits while being more rapid when compared to the conventional method and without compromise of accuracy. The extracted DNA recovery, purity and PCR amplifiability has improved over the conventional method thus can also be recommended as an efficient and feasible method for a more sustainable or routine analysis for EBN identification. With no consideration on cost, the commercial kit, Qiagen method ranked the best in terms of highest DNA purity and PCR amplifiability for DNA sequencing to identify swiftlet species of EBN.

Author Contributions: Conceptualization, M.C.Q. and N.L.C.; methodology, M.C.Q. and S.W.T.; validation, M.C.Q.; formal analysis, M.C.Q.; investigation, M.C.Q.; resources, S.W.T.; data curation, M.C.Q.; writing—original draft preparation, M.C.Q.; writing—review and editing, N.L.C.; visualization, M.C.Q.; supervision, N.L.C. and S.W.T.; project administration, N.L.C.; funding acquisition, N.L.C. All authors have read and agreed to the published version of the manuscript.

Funding: This research was funded by The Ministry of Education, Malaysia, grant number ERGS/1/2013/TK05/UPM/02/6.

Acknowledgments: The authors acknowledge Yusof, Y.A. and Law, C.L. for being supervisory committee member of the first author during her Ph.D. studies.

Conflicts of Interest: The authors declare no conflict of interest.

References

1. Wu, Y.J.; Chen, Y.; Wang, B.; Bai, L.Q.; Han, W.R.; Ge, Y.Q.; Yuan, F. Application of SYBRgreen PCR and 2DGE methods to authenticate edible bird's nest food. *Food Res. Int.* **2010**, *43*, 2020–2026. [CrossRef]
2. Marcone, M.F. Characterization of the edible bird's nest the "Caviar of the East". *Food Res. Int.* **2005**, *38*, 1125–1134. [CrossRef]
3. Guo, L.L.; Wu, Y.J.; Liu, M.C.; Wang, B.; Ge, Y.Q.; Chen, Y. Authentication of edible bird's nests by TaqMan-based real-time PCR. *Food Control* **2014**, *44*, 220–226. [CrossRef]
4. Lin, J.R.; Zhou, H.; Lai, X.P.; Hou, Y.; Xian, X.M.; Chen, J.N.; Wang, P.X.; Zhou, L.; Dong, Y. Genetic identification of edible birds' nest based on mitochondrial DNA sequences. *Food Res. Int.* **2009**, *42*, 1053–1061. [CrossRef]
5. Chapela, M.J.; Sotelo, C.G.; Pérez-Martín, R.I.; Pardo, M.Á.; Pérez-Villareal, B.; Gilardi, P.; Riese, J. Comparison of DNA extraction methods from muscle of canned tuna for species identification. *Food Control* **2007**, *18*, 1211–1215. [CrossRef]
6. Lenstra, J.A. DNA methods for identifying plant and animal species in food. In *Food Authenticity and Traceability*; Lees, M., Ed.; Woodhead Publishing Ltd.: Cambridge, UK, 2003; pp. 34–53.
7. Woolfe, M.; Primrose, S. Food forensics: Using DNA technology to combat misdescription and fraud. *Trends. Biotechnol.* **2004**, *22*, 222–226. [CrossRef]
8. Di Pinto, A.; Forte, V.; Guastadisegni, M.C.; Martino, C.; Schena, F.P.; Tantillo, G. A comparison of DNA extraction methods for food analysis. *Food Control* **2007**, *18*, 76–80. [CrossRef]
9. Karlsson, A.O.; Holmlund, G. Identification of mammal species using species-specific DNA pyrosequencing. *Forensic. Sci. Int.* **2007**, *173*, 16–20. [CrossRef]
10. Schnell, I.B.; Fraser, M.; Willerslev, E.; Gilbert, M.T.P. Characterisation of insect and plant origins using DNA extracted from small volumes of bee honey. *Arthropod. Plant Interact.* **2010**, *4*, 107–116. [CrossRef]
11. Lago, F.C.; Herrero, B.; Madriñán, M.; Vieites, J.M.; Espiñeira, M. Authentication of species in meat products by genetic techniques. *Eur. Food Res. Technol.* **2011**, *232*, 509–515. [CrossRef]
12. Mayer, H.K. Milk species identification in cheese varieties using electrophoretic, chromatographic and PCR techniques. *Int. Dairy J.* **2005**, *15*, 595–604. [CrossRef]
13. Jérôme, M.; Lemaire, C.; Verrez-Bagnis, V.; Etienne, M. Direct sequencing method for species identification of canned sardine and sardine-type products. *J. Agric. Food Chem.* **2003**, *51*, 7326–7632. [CrossRef] [PubMed]
14. Maede, D. A strategy for molecular species detection in meat and meat products by PCR-RFLP and DNA sequencing using mitochondrial and chromosomal genetic sequences. *Eur. Food Res. Technol.* **2006**, *224*, 209–217. [CrossRef]
15. Han, J.X.; Wu, Y.J.; Huang, W.S.; Wang, B.; Sun, C.F.; Ge, Y.Q.; Chen, Y. PCR and DHPLC methods used to detect juice ingredient from 7 fruits. *Food Control* **2012**, *25*, 696–703. [CrossRef]
16. Busconi, M.; Foroni, C.; Corradi, M.; Bongiorni, C.; Cattapan, F.; Fogher, C. DNA extraction from olive oil and its use in the identification of the production cultivar. *Food Chem.* **2003**, *83*, 127–134. [CrossRef]
17. Datukishvili, N.; Gabriadze, I.; Kutateladze, T.; Karseladze, M.; Vishnepolsky, B. Comparative evaluation of DNA extraction methods for food crops. *Int. J. Food Sci. Tech.* **2010**, *45*, 1316–1320. [CrossRef]
18. Rasmussen, R.S.; Morrissey, M.T. DNA-based methods for the identification of commercial fish and seafood species. *Compr Rev Food Sci. Food Saf.* **2008**, *7*, 280–295. [CrossRef]
19. Abdel-Latif, A.; Osman, G. Comparison of three genomic DNA extraction methods to obtain high DNA quality from maize. *Plant Methods* **2017**, *13*, 1. [CrossRef]
20. Corcoll, N.; Osterlund, T.; Sinclair, L.; Eiler, A.; Kristiansson, E.; Backhaus, T.; Martin Eriksson, K. Comparison of four DNA extraction methods for comprehensive assessment of 16S rRNA bacterial diversity in marine biofilms using high-throughput sequencing. *FEMS Microbiol. Lett.* **2017**, *364*, fnx139. [CrossRef]
21. Kek, S.P.; Chin, N.L.; Tan, S.W.; Yusof, Y.A.; Chua, L.S. Comparison of DNA extraction methods for entomological origin identification of honey using simple additive weighting method. *Int. J. Food Sci. Technol.* **2018**, *53*, 1–10. [CrossRef]
22. Pipan, B.; Zupancic, M.; Blatnik, E.; Dolnicar, P.; Meglic, V. Comparison of six genomic DNA extraction methods for molecular downstream applications of apple tree (*Malus X domestica*). *Cogent Food Agric.* **2018**, *4*, 1540094. [CrossRef]

23. Lin, J.R.; Zhou, H.; Lai, X.P.; Hou, Y.; Xian, X.M.; Chen, J.N.; Wang, P.X.; Zhou, L.; Dong, Y. The DNA extraction method of edible bird's nest. *World Sci. Technol. -Mod. Tradit. Chin. Med. Mater. Med.* **2010**, *12*, 202–210.
24. Lee, P.L.M.; Clayton, D.H.; Griffiths, R.; Page, R.D.M. Does behavior reflect phylogeny in swiftlets (Aves: Apodidae)? A test using cytochrome *b* mitochondrial DNA sequences. *Proc. Natl. Acad. Sci. USA* **1996**, *93*, 7091–7096. [CrossRef] [PubMed]
25. Abdullah, L.; Adawiyah, C.W.R. Simple additive weighting methods of multi criteria decision making and applications: A decade review. *Int. J. Inf. Proces. Manag.* **2014**, *5*, 39–49.
26. Gurmeric, V.; Dogan, M.; Toker, O.; Senyigit, E.; Ersoz, N. Application of different multi-criteria decision techniques to determine optimum flavour of prebiotic pudding based on sensory analyses. *Food Bioprocess Technol.* **2013**, *6*, 2844–2859. [CrossRef]
27. Zardari, N.H.; Ahmed, K.; Shirazi, S.M.; Yusop, Z.B. *Weighting Methods and Their Effects on Multi-Criteria Decision Making Model Outcomes in Water Resources Management*; Springer: New York, NY, USA, 2014.
28. Janic, M.; Reggiani, A. An application of the multiple criteria decision making (MCDM) analysis to the selection of a new hub airport. *Eur. J. Transp. Infrast Res.* **2002**, *2*, 113–142.
29. Memariani, A.; Amini, A.; Alinezhad, A. Sensitivity analysis of simple additive weighting method (SAW): The results of change in the weight of one attribute on the final ranking of alternatives. *J. Ind. Eng.* **2009**, *4*, 13–18.
30. Rao, R.V. Introduction to multiple attribute decision-making (MADM) methods. In *Decision Making in the Manufacturing Environment: Using Graph Theory and Fuzzy Multiple Attribute Decision Making Methods*; Rao, R.V., Ed.; Springer Science & Business Media: London, UK, 2007; pp. 27–41.
31. Murray, S.R.; Butler, R.C.; Hardacre, A.K.; Timmerman-Vaughan, G.M. Use of quantitative real-time PCR to estimate maize endogenous DNA degradation after cooking and extrusion or in food products. *J. Agrc. Food Chem.* **2007**, *55*, 2231–2239. [CrossRef] [PubMed]
32. Chen, H.; Rangasamy, M.; Tan, S.Y.; Wang, H.; Siegfried, B.D. Evaluation of five methods for total DNA extraction from western corn rootworm beetles. *PLoS ONE* **2010**, *5*, e11963. [CrossRef] [PubMed]
33. Pirondini, A.; Bonas, U.; Maestri, E.; Visioli, G.; Marmiroli, M.; Marmiroli, N. Yield and amplifiability of different DNA extraction procedures for traceability in the dairy food chain. *Food Control* **2010**, *21*, 663–668. [CrossRef]
34. Cawthorn, D.-M.; Steinman, H.A.; Witthuhn, R.C. Comparative study of different methods for the extraction of DNA from fish species commercially available in South Africa. *Food Control* **2011**, *22*, 231–244. [CrossRef]
35. Besbes, N.; Fattouch, S.; Sadok, S. Comparison of methods in the recovery and amplifiability of DNA from fresh and processed sardine and anchovy muscle tissues. *Food Chem.* **2011**, *129*, 665–671. [CrossRef] [PubMed]
36. Shokere, L.A.; Holden, M.J.; Jenkins, G.R. Comparison of fluorometric and spectrophotometric DNA quantification for real-time quantitative PCR of degraded DNA. *Food Control* **2009**, *20*, 391–401. [CrossRef]
37. Teare, J.M.; Islam, R.; Flanagan, R.; Gallagher, S.; Davies, M.G.; Grabau, C. Measurement of nucleic acid concentrations using the DyNA Quant and the GeneQuant. *Biotechniques* **1997**, *22*, 1170–1174. [CrossRef]
38. Turci, M.; Sardaro, M.L.S.; Visioli, G.; Maestri, E.; Marmiroli, M.; Marmiroli, N. Evaluation of DNA extraction procedures for traceability of various tomato products. *Food Control* **2010**, *21*, 143–149. [CrossRef]
39. Sambrook, J.; Russell, D.W. *Molecular Cloning: A Laboratory Manual*, 3rd ed.; Cold Spring Harbor Laboratory Press: New York, NY, USA, 2001.
40. Mohammadi, T.; Reesink, H.W.; Vandenbroucke-Grauls, C.M.J.E.; Savelkoul, P.H.M. Removal of contaminating DNA from commercial nucleic acid extraction kit reagents. *J. Microbiol. Meth.* **2005**, *61*, 285–288. [CrossRef] [PubMed]
41. Chan, P.K.S.; Chan, D.P.C.; To, K.F.; Yu, M.Y.; Cheung, J.L.K.; Cheng, A.F. Evaluation of extraction methods from paraffin wax embedded tissues for PCR amplification of human and viral DNA. *J. Clin. Pathol.* **2001**, *54*, 401–403. [CrossRef] [PubMed]
42. Willfinger, W.W.; Mackey, K.; Chomczynski, P. Assessing the quantity, purity and integrity of DNA and DNA following nucleic acid purification. In *DNA sequencing II: Optimizing Preparation and Cleanup*; Kieleczawa, J., Ed.; Jones and Bartlett Publishers, Inc.: London, UK, 2006; pp. 291–308.

MDPI
St. Alban-Anlage 66
4052 Basel
Switzerland
www.mdpi.com

Foods Editorial Office
E-mail: foods@mdpi.com
www.mdpi.com/journal/foods

Disclaimer/Publisher's Note: The statements, opinions and data contained in all publications are solely those of the individual author(s) and contributor(s) and not of MDPI and/or the editor(s). MDPI and/or the editor(s) disclaim responsibility for any injury to people or property resulting from any ideas, methods, instructions or products referred to in the content.

www.ingramcontent.com/pod-product-compliance
Lightning Source LLC
LaVergne TN
LVHW070604100526
838202LV00012B/561